PROFESSIONAL ROPE ACCESS

PROFESSIONAL ROPE ACCESS

A Guide to Working Safely at Height

Loui McCurley

WILEY

Copyright © 2016 by John Wiley & Sons, Inc. All rights reserved

Published by John Wiley & Sons, Inc., Hoboken, New Jersey
Published simultaneously in Canada

No part of this publication may be reproduced, stored in a retrieval system, or transmitted in any form or by any means, electronic, mechanical, photocopying, recording, scanning, or otherwise, except as permitted under Section 107 or 108 of the 1976 United States Copyright Act, without either the prior written permission of the Publisher, or authorization through payment of the appropriate per-copy fee to the Copyright Clearance Center, Inc., 222 Rosewood Drive, Danvers, MA 01923, (978) 750-8400, fax (978) 750-4470, or on the web at www.copyright.com. Requests to the Publisher for permission should be addressed to the Permissions Department, John Wiley & Sons, Inc., 111 River Street, Hoboken, NJ 07030, (201) 748-6011, fax (201) 748-6008, or online at http://www.wiley.com/go/permissions.

Limit of Liability/Disclaimer of Warranty: While the publisher and author have used their best efforts in preparing this book, they make no representations or warranties with respect to the accuracy or completeness of the contents of this book and specifically disclaim any implied warranties of merchantability or fitness for a particular purpose. No warranty may be created or extended by sales representatives or written sales materials. The advice and strategies contained herein may not be suitable for your situation. You should consult with a professional where appropriate. Neither the publisher nor author shall be liable for any loss of profit or any other commercial damages, including but not limited to special, incidental, consequential, or other damages.

For general information on our other products and services or for technical support, please contact our Customer Care Department within the United States at (800) 762-2974, outside the United States at (317) 572-3993 or fax (317) 572-4002.

Wiley also publishes its books in a variety of electronic formats. Some content that appears in print may not be available in electronic formats. For more information about Wiley products, visit our web site at www.wiley.com.

Library of Congress Cataloging-in-Publication Data:

Names: McCurley, Loui, 1965- author.
Title: Professional rope access : a guide to working safely at height / Loui
   McCurley.
Description: Hoboken, New Jersey : John Wiley & Sons, Inc., [2016] | Includes
   bibliographical references and index.
Identifiers: LCCN 2016002282 | ISBN 9781118859605 (cloth)
Subjects: LCSH: Falls (Accidents)–Prevention. | Rappelling–Safety measures.
   | Building–Safety measures.
Classification: LCC T55.3.H45 M355 2016 | DDC 363.11/073–dc23 LC record available at http://lccn.loc.gov/2016002282

Typeset in 10/12pt, PalatinoLTStd by SPi Global, Chennai, India

Printed in the United States of America

10 9 8 7 6 5 4 3 2 1

Contents

Notes on Contributors xiii

Foreword xix

Your Success xxiii

Preface xxv

How to use this book xxvii

Section 1 | *Planning for Rope Access* 1

Chapter 1 | *What Is Rope Access?* 3

- 1-1 **Introduction** 3
  - Why Choose Rope Access? 4
- 1-2 **Foundation and Evolution of Rope Access** 6
  - The Modern Rope Access System 6
  - Skills Required for Modern Rope Access Workers 7
- 1-3 **Rope Access Compared/Contrasted with Other Disciplines** 8
  - Rope Access Contrasted with Recreational Rappelling 8
  - Rope Access Contrasted with Controlled Descent 9
  - Rope Access Contrasted with Rope Descent Techniques 11
  - Rope Access Contrasted with Bosun's Chairs 11
  - Rope Access Backup Contrasted with Other Backup Systems 14
  - Where Modern Rope Access Has Landed – The Bus You Take to Work 14
- 1-4 **Compatibility and Work Safety** 15
  - Fall Restraint 15
  - Fall Arrest 15
  - Fall Containment Systems 17
  - Suspended Scaffolds 17
  - Aerial Platforms and Man Baskets 17
- 1-5 **Special Techniques** 18
  - Aid Climbing 18
  - Lead Climbing 19
- 1-6 **Practical Application of Rope Access** 19
- 1-7 **Summary** 20

Chapter 2 | Rope Access and the Comprehensive Managed Fall Protection Plan — 21

- 2-1 Protecting Workers at Height 21
- 2-2 Comprehensive Managed Fall Protection 22
  - Policy Statement 22
  - Staff Responsibilities 23
  - Fall Hazard Survey 25
- 2-3 Hierarchy of Fall Protection 26
  - Types of Active Protection – Harness-Based Solutions 27
  - Choosing a System of Protection 28
- 2-4 Rope Access Work Plan 29
  - System Requirements 30
  - Safety in Rope Access Operations 31
  - Work Practices 32
- 2-5 Summary 34

Chapter 3 | Personnel Selection and Training — 35

- 3-1 Introduction 35
- 3-2 Personnel Qualifications 35
  - Aptitude for Rope Access 37
- 3-3 Team Organization and Competencies 38
  - Technician Skills and Responsibilities 38
  - Supervisor Skills and Responsibilities 39
  - Program Manager Skills and Responsibilities 40
- 3-4 Training and Certification 40
  - Training Records 41
  - Training Outline 41
  - Advanced Levels of Certification 45
  - Rope Access Certification Bodies 46
- 3-5 Summary 47

Chapter 4 | Equipment for Rigging — 49

- 4-1 Equipment for Rigging Rope Access Systems 49
- 4-2 How to Choose Equipment for Rigging in Life Safety Situations 50
- 4-3 The Difference Between Personal Equipment and Rigging Equipment 52
- 4-4 Rigging Equipment for Fall Arrest 52
- 4-5 Rigging Equipment for Cranes Versus Rigging Used in Rope Access 53
- 4-6 Rigging Equipment for Suspended Scaffolds 54
- 4-7 Rigging Equipment for Rescue 54
- 4-8 Rope 55
- 4-9 Connectors 59
- 4-10 Hardware 63
- 4-11 Mechanical Anchorage Connectors 65
- 4-12 Anchor Slings 67
- 4-13 Summary 69

CONTENTS

Chapter 5 | *Personal Equipment for Rope Access*   **71**

    5-1   Introduction 71
    5-2   Essential Requirements 72
    5-3   Harnesses 74
    5-4   A Note About Comfort Seats (Seatboards) 76
    5-5   Helmets 77
    5-6   Lanyards 78
    5-7   Connectors 79
    5-8   Descending Devices 80
    5-9   Rope Access Backup Devices 84
    5-10 Ascenders 86
    5-11 Gloves 88
    5-12 Clothing and Personal Wear 89
    5-13 OTHER PPE 89
    5-14 Equipment Traceability and Recordkeeping 89
    5-15 Summary 90

Section 2 | *Skills for the Rope Access Technician*   **91**

Chapter 6 | *Rigging Concepts*   **93**

    6-1   Principles at Work in a System 94
           Gravity 94
           Friction 94
           Angles 96
           Vector Forces 99
    6-2   Using the Principles 107
           Mechanical Advantage 107
           Load Ratios 111
           Safety Factors 112
    6-3   Summary 114

Chapter 7 | *Rope Terminations and Anchorages*   **115**

    7-1   Rope and Knot Terminology 116
    7-2   Rope Terminations 118
    7-3   Manufactured Terminations 118
    7-4   Knots 119
           Stopper Knots 119
           End-of-Line Knots 120
           Midline Knots 124
           Knots (Bends) That Join Two Ropes 125
           Knot Safety 127
           Hitches 128
    7-5   Knots and Rope Strength 132
    7-6   Anchorages 132
           Classifications of Anchorages 133
           Anchorage System Performance 134
           Positioning the Anchorage System 135

Back-Ties   136
Direct Attachment   136
Load Sharing Anchor Systems   138
Angles in Anchor Systems   141
Change of Direction   142
7-7   Summary   143

Chapter 8 | Rope Access Systems   145

8-1   Compatibility   149
8-2   Access System   149
8-3   Backup System   151
8-4   Attachment to Technician's Harness   152
8-5   Pull-through Systems   153
Pull-Through with Ground Anchor   153
Top Anchor Pull-Through with Knot   153
8-6   Changing the Fall Line   155
Directional Deviation   155
Rebelay (Re-anchor) Systems   156
Well-Being of the Technician   157
8-7   Summary   159

Chapter 9 | Descending   161

9-1   Introduction   161
9-2   Choosing a Descender   162
9-3   Choosing a Rope for Descent   164
9-4   Rigging for Descent   165
9-5   Getting on Rope   167
9-6   Managing the Descent   168
9-7   Tending the Backup Device   170
9-8   Passing a Knot   170
9-9   Passing a Deviation Anchor   173
9-10   Passing a Rebelay   174
9-11   Landing   177
9-12   Summary   178

Chapter 10 | Ascending   179

10-1   Selecting Ascenders   180
Handled Ascenders   180
Chest Ascender   181
10-2   The Complete Ascending System   181
Rigging the Chest Ascender   182
Rigging the Handled Ascender   184
10-3   Managing the Ascent   184
10-4   Changeovers   186
Changeover from Ascending System to Descending System   187
Using a Descender for Ascent   187

CONTENTS

          Rope-to-Rope Transfer 188
          Passing a Knot in the Ropes While on Ascent 189
          Negotiating an Edge or Obstruction While on Ascent 190
          Passing a Deviation 191
          Passing a Re-anchor (Rebelay) on Ascent 192
          Transitioning Off Rope from Ascent, Onto a Platform 196
   10-5 Summary 196

Chapter 11 | *Advanced Techniques*   197

   11-1 Belays 198
   11-2 Aid Climbing 201
   11-3 Lead Climbing 203
   11-4 Climbing with Twin Lanyards 206
   11-5 Raising and Lowering Systems 207
   11-6 Systems for Lowering 208
   11-7 Systems for Raising 209
   11-8 Cross-haul 211
   11-9 Tensioned Ropes 212
   11-10 Multiple Simultaneous Systems 214
   11-11 Powered Assist Systems 214
   11-12 Summary 215

Chapter 12 | *Use of Powered Rope Access Devices*   217

   12-1 Precautions 217
   12-2 Configuring the Device into the System 219
   12-3 Configuration 1 (Sit on Top) 221
   12-4 Configuration (Suspend Beneath) 223
   12-5 Using the Device from a Fixed Position 224
   12-6 Additional Considerations 226
   12-7 Care and Maintenance 226
   12-8 Summary 227

Chapter 13 | *Rescue*   229

   13-1 Rope Access and Rescue 229
   13-2 The Rescue Preplan 232
   13-3 Self-Rescue 233
   13-4 Coworker-Assisted Rescue 235
   13-5 Noncommittal Rescue and Prerigging for Rescue 236
   13-6 Co-Worker Assisted Rescue from Descent 238
   13-7 Rescue from Ascent 240
   13-8 Challenging Rescues 241
   13-9 Standby Rescue 242
   13-10 Professional Versus Coworker-Assisted Rescue 243
   13-11 Conclusion 244

Section 3 | *Program Administration*   **245**

Chapter 14 | *Developing a Rope Access Plan*   **247**

14-1 Working Safely at Heights   248
     Harness-Based Works   248
     Fall Arrest   250
     Work Positioning   250
     Rope Access   251
14-2 Necessary Elements of a Rope Access Program   251
     Rescue   251
     One Rope or Two?   252
     Avoid the Fall   253
     Team Works   253
     Team Documentation   254
14-3 Work in a System   254
     Supervisor   255
     Suitable Management   255
Summary   256

Chapter 15 | *Developing a Policy Statement*   **257**

15-1 Questions to Consider   257
15-2 Putting it all Together   262
15-3 Congratulations!   263

Chapter 16 | *Writing a Work Order*   **265**

Summary   268

Chapter 17 | *Establishing a Work Plan*   **269**

17-1 Summary   273

Chapter 18 | *Performing a Job Hazard Analysis*   **275**

18-1 The Process   275
18-2 Content   276
18-3 Using the JHA   278
18-4 Summary   279

Chapter 19 | *Fall Hazard Survey/Assessment*   **281**

19-1 Conducting the Survey   281
19-2 Survey Contents   283
19-3 Using the Survey   283
19-4 Fall Hazard Mitigation   284
19-5 Summary   284

Chapter 20 | Creating a Rescue Preplan — 287

- 20-1 Emergency Response Planning 287
- 20-2 Fall Rescue Planning 289
- 20-3 Coordination with External Resources 293
- 20-4 Summary 293

Chapter 21 | Training Records — 295

- 21-1 Certification Records 295
- 21-2 Technician Records 296
- 21-3 Employer Records 297
- 21-4 Program Administrator Training 298

Chapter 22 | Equipment Inspection and Care — 301

- 22-1 Specifying Equipment 301
- 22-2 Placing Equipment into Service 302
- 22-3 Equipment Inspection 302
- 22-4 Cleaning 303
- 22-5 Retirement 304

Chapter 23 | Rope Access Program Audit — 307

- 23-1 Components of an Audit 307
- 23-2 Management 308
- 23-3 Practices 309
- 23-4 Equipment Management 311
- 23-5 Recordkeeping 312
- 23-6 Summary 312

Knowledge Check 313

Glossary 339

Index 343

In memory of Steve Hudson

Steve Hudson

Notes on Contributors

Although my name is on the jacket, I can only take so much credit for this work.

Isaac Newton wrote in the year 1676 *"If I have seen further, it is by standing on the shoulders of giants,"* and I feel no less small as I consider the pool of knowledge that has contributed so greatly to this text. From the men who employed rope access methods (without calling it that) during the building of the Hoover Dam in the 1930s to the myriad of window cleaners, bosun's mates and chimney sweeps who employed ropes in their respective trades, and the pioneering work of IRATA in the late 1900s; the techniques and equipment employed by modern rope access technicians would be nothing without these.

I would especially like to recognize the significant contributions of my dear friend and height safety magnate, the late Steve Hudson in the development of this work – and indeed the world of work at height in general.

In the mid 1990s, when I was lamenting about the absence of an open-forum, democratic body in which to pool knowledge, experience, and expertise toward the improvement of rope access safety, practices, and understanding, Steve's response was *"well, why don't you do something about it?"*

With this as a backdrop, Steve proceeded to guide and facilitate my passion, using his knowledge and the resources at Pigeon Mountain Industries, Inc., to underwrite the development of the Society of Professional Rope Access Technicians (SPRAT). The two of us, together with James Frank of CMC Rescue, Inc., and Michael Roop of Roco Rescue, Inc., served as the founding Board of Directors of SPRAT, nurturing and prodding the practice of rope access through its infancy in the United States and beyond.

If there ever was a giant upon whose shoulders this industry was built, it was Steve. Indeed, if not for Steve's encouragement and belief in me I don't suppose I would ever have had anything to do with rope access at all.

This book also would not be what it is without the contributions and knowledge of the industry professionals who lent their depth and experience through the contribution of chapters. My heartfelt thanks goes to these individuals.

TOM WOOD, CHAPTERS 4 AND 13

Tom Wood

Tom is the Training Manager for PMI's Vertical Rescue Solutions, and a SPRAT certified Level 3 Supervisor. Actively involved in mountain rescue, he is also a member of the Alpine Rescue Team in Evergreen, Colorado, and a Mountain Rescue Association (MRA) Terrestrial Delegate to the International Commission for Alpine Rescue (ICAR). He lives in Conifer, Colorado, with his wife and three children.

KEITH LUSCINSKI, CHAPTER 9

Keith Luscinski

Keith loves all things rope related. He became a work-at-height professional while developing courses in tree climbing and technical rock climbing rescue for Cornell University, where he studied Operations Research and Industrial Engineering. As an industrial rope access technician, Keith has found himself on the Gateway Arch and the Empire State Building as well as myriad other buildings, bridges, and structures.

PETER FERGUSON, CHAPTER 14

Peter Ferguson

Peter has been a leading figure and consultant in Australia's fall protection, pedestrian access, rope access, and suspended access industries for more than 40 years. Founder of the Australian Rope Access Association (ARAA), Peter convened and was the driving force behind ISO 22846 Rope Access Standard. Recipient of the Workssafe Victoria award and Standards Australia award, Peter serves on the board of directors of the International Society for Fall Protection.

The success of continuity and detail of this work is directly attributable to the work of a few "behind the scenes" individuals – knowledgeable and committed content editors who have spent countless hours poring over copy, contributing gallons of red ink to the project, and inserting suggestions and ideas to make the text more accurate and usable for safety professionals and technicians:

BOB MCCURLEY

Combined with a 20 plus year career in physics and engineering, Bob is a SPRAT certified rope access technician, member of the Alpine Rescue Team in Colorado, past volunteer firefighter, and part-time instructor for industrial rescue.

ROB DUNSHEA

Rob has worked in rope access for 25 years across industrial, military, rescue, and training domains. He is Director of Assessment for the ARAA and sits on the SPRAT Evaluations Committee.

THOMAS EVANS

Tom is cofounder of a rigging research and teaching nonprofit (SAR^3), avid caver, and instructor with the National Cave Rescue Commission. He is passionate about evidence-based decision making in rigging.

If the words of this text are important, so are the images. The following people were instrumental in helping to illustrate the concepts outlined in this text through the contribution of photographs and images. Without their talents, the limitations of the words would be glaring.

- Margaret DeLuca (Illustrations)
- Trask Bradbury, Gemini
- Trask Bradbury, Gemini
- Ken Piposar, Abseilon
- Karl Guthrie, ClimbTech
- Chris Vinson, ClimbTech
- Dan Henn, Reliance Industries LLC
- Ken Hauser, PMI
- TomWood, PMI's Vertical Rescue Solutions
- Bob McCurley, PMI
- Sean Cogan, Harken Industrial
- Beal/Vuedici
- Don Enos, SMC
- Michel Goulet, Petzl
- Jody Bird, SPRAT
- Robert Gray, Transystems
- Michael Seal, Burgess & Niple
- Tractel
- Jay Smith

And, finally, to those who – knowingly or otherwise – have contributed to my education and knowledge through the years. To paraphrase my dear friend Scott Mohon, "knowledge doesn't belong to anyone. It exists to be shared." So, to all who have shared your knowledge over the years: thank you! This text would be sorely lacking without the guidance and coaching you've provided over the past several decades. You are too many to mention, but you know who you are. I thank you from the bottom of my heart.

Foreword

A worker dies from a fall every single work day in America. Appropriately, fall protection violations are the most cited construction activity by OSHA. As a safety consultant, I am obligated to review national worker fatality reports looking for causes and even better ways to protect my clients. The facts are clear to me – most fall deaths are due to poor planning and lack of preparation for the elevated work undertaken. Certainly, bad equipment choices (read techniques) and insufficient training are major contributing factors contained under that planning umbrella.

Of all the thousands of workers that have died from falls over the past 10 years, none of them planned to die that day. Most of them had solid footing under them – the infamous "false sense of security" which can lead to workers taking unnecessary risks. What gets my attention in the reports is the number of folks killed from falls from rooftops, ladders, and scaffolding which OSHA has targeted in its fall protection emphasis. There is surely a better way to do these jobs.

That's why this book is SO IMPORTANT to anybody that must put workers in "elevation jeopardy." The primary author, Loui McCurley, is an experienced and knowledgeable voice telling you that there is a viable alternative to elevated work taking place on walking working surfaces that workers can, and do, fall from. I have had the honor of working with Loui on fall protection device testing and technique development, the ANSI Z359 Fall Protection safety standard committee, as well as the National Fire Protection Association (NFPA) rescue standards body. Loui and I are also among the cofounders of an organization called the Society of Professional Rope Access Technicians (SPRAT), which developed standards and certifications for rope access work and where Loui still serves as the Regulatory Assistance Chairman. So, as you read this text, be assured that the author can not only capably talk the talk that she has penned here – she has and does walk the walk.

Many workplaces have the equipment and ability to arrest a fall (or what I like to call the FIRST rescue), but invariably don't have a good plan to facilitate the fallen worker's rescue (which is the SECOND rescue). The fallen worker is often left suspended for an inordinate amount of time while his coworkers attempt to make a save they are, often as not, ill prepared to make happen. As minutes tick by, the fallen worker is now in danger of making a panicky mistake (such as trying to get out of his harness because he's thinking "I'm only 15 feet or so from the ground"), not to mention the deadly peril he or she is in due to harness suspension trauma. Yes, it is indeed my intent to scare you – because you need to pay attention to this book!

Full disclosure, you may have guessed from the above paragraphs that my background is emergency response and I therefore tend to look at fall protection from the prospective of the fallen worker (fall victim). This book is valuable for the variety of answers it provides to manage the fall menace. Crucially, the book posits a solution to making that SECOND rescue... instead of a worker becoming a victim dangling at the end of a synthetic line (fall protection lanyard) after a fall; why not consider starting the work out hanging on a synthetic line (rope) to get the job done? The fallen worker does not have to be a fall victim with little control over their fate... they can be the rope access worker with complete control of their safe work. My personal experience has shown me time and again that utilizing rope access to get many elevated jobs done is a safe, efficient process that allows the worker to easily get him or herself out of sticky situations if they arise. Especially when utilizing ladders and scaffolds, rope access is a much better (more efficient) work method while providing safer task management. To draw from another field from which I have some experience, confined space work – when it becomes necessary, the best rescue is self-rescue and rope access provides the opportunity for workers to readily save themselves!

Let me explain the differences. Workers tend to have some sense of security when standing on what seems to be a solid surface such as a ladder or scaffold floor – especially if they are wearing a fall protection harness and lanyard. So, the task becomes their primary concern and the height from which they are working is not as important as getting the job done. The reality of a dangerous elevated work location diminishes as the worker's anxiety lessens and they feel more at ease. Safety is relegated to the back of the mind. You've heard it before, but I'm going to say it again – a little fear breeds respect for the hazard(s) that confront you. If a fall occurs and the worker happens to survive the fall arrest, she/he faces a shocking reality – they almost just died from falling and now have NO PLATFORM under their feet. Falling is now a reality to them and they are left hanging wondering, "What's next?!" Any normal person's mind is going to reel from what they just experienced plus their current hazardous predicament. "All aboard! Next stop – PANIC!" Of course, not every fallen worker will panic, which causes loss of rationale/sensibility, but some people don't handle stress well.

Conversely, if this type of job was done via rope access, workers are constantly reminded of where they are and how their lives depend on keeping safety on an equal level with job efficiency. I have seen firsthand how workers who are confident in their own role in their personal safety are more competent (effective and efficient) in their work. Well-trained rope access workers get their job done by staying focused first on their safety rigging (main line and backup safety line), which allows them to then expedite the task. Again, they are always aware of where they are and what they are doing – nothing is allocated to "muscle memory" or a monotonous repetitive task... nothing is taken for granted. Just between you and me, rope access allows the worker to actually enjoy the work while getting it done in a much timelier manner at often a much reduced cost.

I spend an enormous amount of time developing, training, and auditing my clients' fall protection programs to not only comply with regulatory requirements, but more importantly, to actually achieve a safe, workable solution to dealing with their fall issues. However, like most fall protection program trainers, my biggest challenge is always to get my clients to focus on that SECOND rescue. This book is written with help from rope access professionals that have addressed strategies and tactics for staying out of trouble at the outset, but if necessary, it presents to you the insight needed for good outcomes.

So, I know that you will enjoy this book. It may be an eye-opening experience for you. Or, maybe, serve as an excellent reference for your fall protection safety library. I'm hoping that the book will convince you that rope access is indeed the best work procedure for many of your elevated tasks. Rope access should not replace your overall fall protection program, but it clearly is a logical, safe, and efficient fall protection tool that should be part of your program.

Michael Roop, CSP

FOREWORD

Michael Roop is a Certified Safety Professional (CSP). His book, Confined Space and Structural Rope Rescue, is available through Mosby/Elsevier Bookstore. Roop is the founder of Roco Rescue, Inc., a rope rescue training and stand-by rescue provider. He is a litigation support expert in fall protection and confined space work. He is also a cofounder of the Society of Professional Rope Access Technicians (SPRAT). Mike is a retired Louisiana State Police Captain.

Your Success

It is my deep desire that this book aids you along the way toward greater safety in your workplace environment, whether you are a worker, a manager, a supervisor, an employee, or an employer. If you have suggestions or recommendations for future revisions of this book, or of companion works, I welcome you to share those with me as I seek to continue in my mission to equip workers at height with tools for safety.

Stay Safe,
Loui McCurley

The Sovereign LORD is my strength! He makes me as surefooted as a deer, able to tread upon the heights. Habakkuk 3:19 (NLT)

Preface

Year after year, falls rank among the top causes of work-related death. Often, the reason given for the worker not being adequately protected is that "protection was not feasible."

This book offers a revolutionary alternative to conventional fall protection methods: Rope Access.

Rope Access provides access and protection for working at heights in environments ranging from the most simple – such as buildings and structures – to the more complex, such as wind turbines and oil platforms. The safety advantages of rope access are grounded in the concepts of 100% tie-off, dual protection, and thoroughly trained workers. Rope Access Technicians ply their skills to access difficult locations to carry out work, often with minimal impact on delicate structures or other operations, while ensuring their safety. With fewer total man-hours and the reduced level of risk, man-at-risk hours can be significantly reduced when compared with other means of access and their associated risks and costs.

Although rope access has been successfully used worldwide for over 20 years with outstanding safety results, it remains poorly understood. This book is the first resource of its kind to attempt to improve awareness and understanding among stakeholders at all levels.

Loui McCurley
Denver, USA
December 2015

How To Use This Book

Whether you are an aspiring rope access technician, a technical safety manager, or an executive with workers at height, the information in this book will help you to be safer, more efficient, and better equipped to do your job.

Executives and safety managers will find this text a useful reference in establishing, overseeing, and maintaining a rope access program, focusing perhaps on Sections 1 and 3, while trainers and practitioners will find it beneficial as a training aid, with emphasis on Sections 1 and 2. While no textbook can ever be a replacement for hands-on, practical experience, this book provides supportive text to augment the training and continuous skills development of rope access technicians and others involved in the rope access safety chain.

Some users of this book will choose to read it cover-to-cover, while others may pick and choose chapters according to their respective needs. The book is organized into chapters, each of which addresses a particular area of rope access. While each chapter stands alone in the area of its content, concepts in some chapters do build and expand upon concepts introduced in other chapters.

For ease of reference, the 23 chapters are grouped into three different sections. Each of the three sections contains related topics.

SECTION 1 – PLANNING FOR ROPE ACCESS

This section is a must-read for Safety Professionals, Regulatory Authorities, and Supervisory Managers. It provides an overview of rope access as a work method, describes rope access within the context of a Comprehensive Managed Fall Protection Program, and provides guidance toward developing the rope access program. It contains five chapters:

CH 1 What Rope Access Is . . . and Isn't!

This chapter provides some historical context for rope access, differentiating it from other suspended rope methods and recreational uses of ropes. Reading this chapter will provide a greater understanding of where rope access concepts have originated, and an informed outlook toward the future.

CH 2 Rope Access and Your Comprehensive Managed Fall Protection Plan

Here we will explore how rope access fits into the Comprehensive Managed Fall Protection Plan, and further develop understanding of the differences between Rope Access and the conventional fall protection methods of Fall Arrest, Positioning, and Restraint.

CH 3 Personnel

In Chapter 3, the human element of rope access is explored, including a review of aptitude, abilities, training, certification, and how to help technicians maintain skills over time. The personnel side of selection and composition of work teams is also addressed, along with team organization and leadership.

CH 4 Equipment for Rigging

In this chapter, we explore the equipment that comprises the systems that rope access technicians use: Artificial Anchors, Rope, Slings and Webbing, Hardware (connectors, rope adjusters, etc.), and more. This encompasses the equipment usually provided by the jobsite or employer for use by a group or team of rope access technicians.

CH 5 Personal Equipment

This chapter provides an overview of personal equipment for rope access, including those items that would be considered Personal Protective Equipment (PPE). It also provides descriptions of each component's use. This includes equipment that is often based on the preference of an individual rope access technician, and that will most often be a part of his personal kit. Selection of equipment is addressed, and PPE for rope access is contrasted with PPE used for other purposes.

SECTION 2 SKILLS FOR THE ROPE ACCESS TECHNICIAN

Having established an understanding of regulations and equipment in Section 1, Section 2 delves into how the equipment is used in the field. Section 2 is a must-read for technicians, as it provides a basis for understanding skills and techniques that are best learned through hands-on training. Administrators and Supervisors will also find this section useful for gaining insight into the safe practices used by the technicians within their area of work.

CH 6 Rigging Concepts

Chapter 6 expands our understanding of the equipment by introducing the factors that affect equipment once it is rigged. Basic concepts of forces and friction are presented, and tips are provided for maximizing the efficiency and effectiveness of the rigging within rope-based systems.

CH 7 Terminations and Anchorages

Appropriately terminating ropes is foundational to every rope access system, as is the implementation within a rope access system of safe, effective anchorages. Because these topics are such an important part of the system, an entire chapter is dedicated to the understanding of termination types and limitations, anchorage terminology, and the types and uses of anchorages.

CH 8 Rope Access Systems

This chapter draws together the information in previous chapters to provide an understanding of how everything fits together as a system for rope access. It offers general guidelines for rigging the basic systems used for moving up and down, and working while being suspended from ropes. A solid understanding of equipment, forces, and rigging concepts is essential to using this chapter effectively.

CH 9 Descending

Chapter 9 begins the foray into the methods that a rope access technician uses as part of their daily work. This chapter discusses important rigging tips for descending systems, methods for managing equipment while descending, negotiating edges, maintaining control, and negotiating basic obstacles including knots, deviations, and re-anchors.

CH 10 Ascending

A companion chapter to Chapter 9, and embracing the idea that any technician who is able to descend down a rope should also be capable of ascending, here the concept of using equipment to ascend a rope is introduced. Rigging tips, equipment management, and special circumstances are all addressed, this time within the context of moving up rather than down.

CH 11 Advanced Techniques

Although much of rope access involves simple up and down movement, the fundamental principle behind the safety of the rope access technician depends upon the ability to perform more advanced techniques when necessary. This includes the safe use of horizontal traverse, lead climbing, aid climbing, twin lanyard climbing, highlines, and guideline systems.

CH 12 Powered Devices

In recent years, the efficiency and safety of rope access has been enhanced through the introduction of powered devices for rope access. Powered devices are not a replacement for skills and training in manual rope access techniques, but rather a tool to assist the technician reduce fatigue and increase efficiency.

CH 13 Rescue

Rope access technicians are capable of safely accessing some of the most remote and otherwise inaccessible work-at-height locations. In fact, rope access technicians are often able to get to locations that professional rescuers are not adequately prepared to reach! For this reason, the ability to perform self-rescue as well as coworker-assisted rescue is a mandatory requirement for all rope access workers. This chapter delves into some detail on this subject.

SECTION 3 PROGRAM ADMINISTRATION

The third and final section of the book is designed to guide the safety manager and/or rope access supervisor through the process of effectively administering a rope access program. Each chapter delves into a different planning tool, some of which are applicable program-wide while others are job-specific. These chapters provide step-by-step guidance to assist administrators through the planning process. Field technicians should also be aware of this information so that they better understand the resources available to them, and how to reference essential job information.

CH 14 Management and Planning for Rope Access

In this chapter, the reader will be reminded of the framework necessary for an effective rope access program and how to establish and adhere to basic principles for a rope access program.

CH 15 Policy Statement (Document Guide)

The employer's policy statement forms the foundation for the comprehensively managed fall protection program, setting the tone for the overall approach to work at height. This chapter will help guide the employer in writing a policy statement.

CH 16 Work Orders (Document Guide)

A work order may be used to help define the scope of work between an organization and a contractor, or between departments within an organization. This chapter provides guidance toward the effective writing and use of work orders.

CH 17 Rope Access Work Plan (Document Guide)

Also sometimes called a "Rope Access Permit," the work plan outlines specific details regarding how a rope access job will be performed, what methods and equipment will be used, personnel responsibilities, and other essential information. This chapter offers insight and assistance toward developing such a document.

CH 18 Job Hazard Analysis (Document Guide)

Where conditions or circumstances exist that might expose a worker to potential hazards in the workplace, a job hazard analysis should be performed to identify and seek mitigation of foreseeable hazards. While not specific to rope access, a JHA is essential in any work at height program. A process for approaching and writing a JHA is offered in this chapter.

CH 19 Fall Hazard Survey (Document Guide)

Every fall protection program begins with a fall hazard survey. While the mission of the job hazard analysis is to identify all potential hazards, the Fall Hazard Survey delves more deeply into hazards associated specifically with potential falls. This chapter provides a process for developing such a document.

CH 20 Rescue Preplan (Document Guide)

Whenever employees are engaged in work at height, the employer should have in place a pre-established plan for prompt rescue in case of a fall. While entire books have been written on this subject, this chapter will offer a synopsis of rescue plan requirements and an approach to developing a plan.

CH 21 Training Records (Document Guide)

The old adage *"if it isn't written down, it didn't happen"* is the reason for this chapter, which emphasizes the importance of training records for both the employer and the employee. This chapter offers insight and provides resources to assist in recordkeeping that is relative to training.

CH 22 Equipment Inspection (Document Guide)

Another area where a systematic approach and good records are essential, equipment inspection is an ever-important part of the gear intensive practice of rope access. In this chapter, a process and system is proposed for equipment inspection and related recordkeeping.

CH 23 Program Audit (Document Guide)

Periodic auditing will help to ensure consistency and excellence in a rope access program. While various rope access organizations offer audit programs specific to their respective requirements, this chapter offers a broad overview of what an audit should contain and how an effective audit might be performed by internal or external resources.

WORKING SAFELY AT HEIGHT

Rope access offers a safe and effective approach to working at height. Equipping and empowering workers to accept responsibility for their own safety when working at height has long been my passion, and I sincerely hope that this book will aid many in that endeavor. Whether you are a worker, a manager, a supervisor, an employee, or an employer, your personal safety is paramount. I applaud you for owning it, and thank you for considering this book as a tool in the process.

SECTION 1

Planning for Rope Access

CHAPTER 1

What Is Rope Access?

By the end of this chapter you should understand:

- How rope access differs from controlled descent, bosun's chairs, rope descent systems, and other aspirant rope based systems
- Some distinct benefits offered by rope access
- Some capabilities of rope access technicians
- The historical context of rope access
- How to differentiate rope access from recreational rope systems
- The essential components of a rope access system
- The importance of a sternal attachment for safety
- Some examples of work applications where rope access is used
- How rope access can interface with other methods of fall protection
- How to approach the implementation of a rope access program.

1-1 INTRODUCTION

Employers around the globe are charged with a complex task of safely managing safety at height in a wide range of activities and industries, and they are expected to do so in a manner that is effective in terms of both safety and fiscal responsibility. At no point in history has this been more critical to the progression of our society than in this post-global recession. Companies and entire industries as a whole are actively working to find new ways to work more efficiently to make more progress with less (workers, money, time, resources, etc.)

Rope access is a specialized mode of access and protection that site owners and managers frequently turn to as a solution when specially trained, certified technicians have to be deployed to hard-to-reach places with maximum safety and minimal cost.

Professional Rope Access: A Guide To Working Safely at Height, First Edition. Loui McCurley.
© 2016 John Wiley & Sons, Inc. Published 2016 by John Wiley & Sons, Inc.

When workers are engaged in working at height, rope is commonly used as a vertical lifeline to provide fall arrest, thereby preventing catastrophic injury in the event of a fall. Rope access takes this approach to safety one step further, by employing two ropes for the safety of the worker: one rope to support the human load and another independently anchored rope for secondary safety.

The secondary safety used in a rope access system is much more conservative than a typical fall arrest system, limiting both fall distance and transmitted-energy potential to very low levels. Rope access is a method of access that provides the user with the means to safely gain access, be supported, as well as a means of egress from a high place, for the purpose of carrying out work.

The term "rope access" encompasses a fairly broad range of capabilities, but properly used rope access is distinctly unique and stands apart from such concepts known as "controlled descent", "rappelling", "bosun's-chairs", "rope descent techniques" and other colloquial terms that have been at times used to generically describe rope-based methods of working at height.

Rope access is unlike any of these, but is unique in that it provides a complete system of access and safety wherein a properly trained and equipped technician will use a completely interchangeable two-rope system. One rope in the system is designed as the primary or working rope, and the other rope is designated as the backup or safety in the system. In a properly rigged rope access setup, as shown in Figure 1-1, each line is fully capable of serving the role of the other. Specifically, the primary line can be used interchangeably to perform the function of the safety line, and vice versa. This interchangeability is essential to achieve the wide variety of capabilities and depth of safety that is unique to rope access.

In the rope access system, the primary/working rope in the system is used not only to ascend or descend a rope, but also to perform a range of movements including passing intermediate anchors, moving horizontally through a rebelay or rope interchange, performing emergency escape, rescuing a coworker, and more. The backup rope serves as just that – a safety backup in case the primary rope system should become compromised or even fail.

Why Choose Rope Access?

For many trades, rope access provides an excellent solution for gaining access to difficult-to-reach locations. The extensive amount of knowledge and training that rope access technicians possess, and the highly disciplined and structured system that rope access comprises mean that more complex jobs can often be performed more safely and efficiently, in a shorter period of time, resulting in both monetary and safety benefits.

For example, when the Arizona Department of Transportation wanted to remove a bird nest from the underside of a bridge on Interstate 10 near Tucson, the inaccessible location of the nest posed an extraordinary challenge. With no walking surfaces within reach of the nest, access was not easily achievable. A heavy duty bucket truck with an articulating arm, known as a "snooper", was initially believed to be the only possible solution, at an estimated cost of over $22,000. Aside from the extraordinary cost, this option would have required lane closures during the work, and would have put workers at risk, in an exposed location, without any back-up alternative for rescue.

After some deliberation, rope access experts were consulted and a system was designed to position suspended workers within easy reach of the nest with 100% backup safety at all times. As an added bonus, the workers were completely

FIGURE 1-1
Rope access requires two independent, interchangeable systems: one for primary access and the other for backup © vuedici.org/BEAL.

self-sufficient and capable of self-rescue and coworker-assisted rescue in case of an emergency – which of course never became necessary. The work was accomplished in less than 4 hours at a cost of less than $4000.[1]

While other methods are sometimes available for protecting workers at height, rope access is an especially good choice for temporary access to locations that are difficult to reach, where other methods might cause physical damage or interference to the structure, where the installation and use of other methods would be time consuming or overly expensive, and also where speed or versatility is desired.

For some trades rope access methods can help get a job done more quickly, efficiently, and safely, than would be possible using conventional fall protection methods. Because the techniques and equipment are conducive to low-impact methods, work can be performed without damage to historical or fragile environments. Rope access workers also have less time exposure to heights, thereby increasing safety.

To realize these advantages to their fullest, strict adherence to the principles of professional rope access is paramount. Professional rope access requires more than just a person with a rope and a descender. It is a complete system of work, with

---

[1] Actual example, submitted by Ken Piposar, Abseilon Rope Access; Davidson Canyon Bridge Project; November 2014.

specific requirements for effective management and application from start to finish. Anything less than this is not rope access, and will not result in the monetary and safety benefits that exemplify true rope access.

1-2 FOUNDATION AND EVOLUTION OF ROPE ACCESS

The early foundations of the rope access system may be found in mountaineering and particularly in caving, where methods and techniques were developed over centuries for traversing dangerous terrain and negotiating vertical spaces. However, the equipment and techniques employed by today's rope access technician bear little resemblance to the equipment used for recreational methods. For the purpose of professional work at height, rope access methods and techniques have been adapted to incorporate specially designed equipment rigged into two separate systems, as described above, with one as the primary support and the other as the safety backup. In contrast, recreational pursuits at height generally rely on only one rope system rather than two. In addition, some components of equipment used for recreation are insufficient for rope access and have been modified or replaced to improve safety.

While rope access today still offers practitioners a means to access the most complex and challenging at-height work spaces, the modernization of rope access has resulted in a system that offers an extraordinary level of protection to practitioners, which exceeds typical fall arrest practices. Given the progress in the adaptations of the methods, and the techniques and equipment used for modern rope access, the resultant professional approach bears only very superficial similarities to its recreational predecessors.

The Modern Rope Access System

In a typical modern rope access system, one rope (the "primary" or "access" line) is used for access and egress and for support at the work location. The worker wears a harness, which is in turn connected to the access line with specially designed devices to accommodate movement and provide support on the rope. The other rope (the "backup" or "secondary" line) is also connected to the user's harness, which is in turn connected to a device that is towed up or down along the safety line as the user ascends or descends the working line. In the event of a failure of the working line or any of its components, the safety system will engage to catch the fall and limit the resulting forces on the user and the equipment. In a properly rigged rope access system, such as that shown in Figure 1-2, the fall protection that the backup line affords is completely separate and independent from the primary means of support.

The two individual rope systems in a complete rope access configuration are at once independent yet interchangeable, offering great versatility while maintaining continuous redundancy.

In addition to having been trained in basic maneuvers for moving up and down the rope, competent rope access technicians are also trained in specialized access methods, such as traversing, aid climbing, and even lead climbing, and they are also trained in more conventional fall arrest and restraint methods as well. This breadth and depth of capability is what permits the rope access worker to protect themselves in such a wide range of challenging environments.

FIGURE 1-2
Backup must be completely separate and independent from the primary means of support. Courtesy of Abseilon USA, www.abseilon.com.

Skills Required for Modern Rope Access Workers

The safe use of rope access and associated systems requires a certain amount of competence on the part of the practitioner. Competence requires not just the capability to perform a skill, but also the ability to understand the logic behind the maneuvers, and to take corrective action when something doesn't go as planned. To this end, every rope access technician should thoroughly understand the reasoning behind the actions they are able to perform. In addition, rope access technicians must be capable of getting themselves out of any predicament they might get themselves into. This requires a range of skills. For example, a technician who can descend must also be capable of ascending; a technician who can pass a knot on ascent must be capable of passing a knot on descent; a technician who can negotiate a deviation in one direction must also be able to negotiate it in the other. Finally, a rope access technician must be capable of performing coworker-assisted rescue in the event that a workmate becomes incapacitated. These techniques, and more, will be addressed in this book.

Gone are the days when just being a climber, or just knowing how to rappel, was sufficient for a rope-based job. The drive toward greater safety and protection for the worker demands competence at a much higher level. Such competence is best acquired through expert training and experience, and should always be confirmed through independent assessment and certification by an appropriate certifying body. The quality of a certifying body will, at least to some extent, dictate the quality of the technicians they endorse.

1-3 ROPE ACCESS COMPARED/CONTRASTED WITH OTHER DISCIPLINES

Rope Access Contrasted with Recreational Rappelling

Some readers may more readily recognize the reference to rope descent by its recreational counterpart, rappelling. This term is used to describe the method commonly used by climbers and cavers to lower themselves from a high point to a low point during mountain climbing or caving activities.

To rappel, rappellers select an anchor (or sometimes two), attach their rope to it, place a rappel device on the rope, attach it to their harness with a carabiner, and back over the edge, as shown in Figure 1-3.

To the casual observer, this concept may seem pretty similar to the method a rope access technician uses to descend rope. However, a close examination and comparison of the equipment and how it is rigged in Figure 1-1, as compared with Figure 1-3, can be enlightening.

Although rope access descent is addressed in detail in Chapter 10, it is imperative here to understand the notable and distinct differences that set professional rope access equipment and methods apart from recreational rappelling.

FIGURE 1-3
A typical recreational rappelling system. Courtesy of Jay Smith.

Foundationally, the first difference is in the rope itself. Rope access technicians select a rope that meets appropriate industry standards, and is rated to a minimum strength of at least 6000 lbf. The rope access rope typically offers very low elongation as compared with the rope most often selected by climbers. This rope is terminated with manufactured terminations (such as sewn or swaged ends) or is "field terminated" by the user using accepted knots that will not reduce the strength of the rope beneath 5000 lbf – a concept explored in more detail in Chapter 7. There is no such requirement for recreational rappelling equipment, and the performance properties of rope used for recreational rappelling may vary widely from that used for rope access.

This differentiation extends to other components of equipment as well. The rope access technician will use locking connectors designed for professional use, whereas the recreational user may use a variety of equipment with performance requirements that are more suited to recreational use. You can read more about equipment for rope access in Chapters 4 and 5.

Perhaps the greatest distinctions between rope access descent techniques and recreational rappelling are specific to the methods used. Specifically, a single rope typically comprises the recreational rappeller's only connection to terra firma. Sometimes called "single rope technique" (SRT), this approach may be appropriate for a recreational climber or caver, but not normally for rope access. The rope access technician is more likely to use two such ropes, as described previously and as shown in Figure 1-1, one as a primary system and the other as backup.

Rope Access Contrasted with Controlled Descent

"Controlled Descent" is a term that evolved into use, primarily by certain industries and related regulatory bodies, to describe the concept of a worker using a recreational rappelling type system as described in the previous section, with conventional fall protection methods (described later in this chapter) as backup. An example of such a system is shown in Figure 1-4. These methods were introduced by window cleaners, chimney sweeps, and others – many of whom were recreational climbers or cavers themselves – who understood the advantages offered by rappelling, and realized that by adding a conventional fall protection system they could be compliant with the regulatory requirements.

While arguably "compliant", there are some distinct concerns with this approach. First and foremost is the fact that the two systems are completely dissimilar and are not designed to function harmoniously with one another. For example, harnesses used for fall arrest are generally designed in an H configuration (Figure 1-5a) and most often offer only a dorsal attachment, whereas harnesses used for rope descent (Figure 1-5b) are designed to hold the wearer in a seated position with a sternal or waist attachment. One outcome of these differences is the fact that the system used for controlled descent is not generally conducive to permitting the worker to ascend out of whatever circumstance he might find himself in.

In addition, the ropes historically used for conventional fall arrest differ from those used for rope access. Fall arrest ropes are most often of a laid variety and incorporate fall arresters that are not easily manipulated on and off the rope, whereas ropes used for rope access are of a kernmantle design and make use of backup devices that accommodate ready transition and transfer. The differences abound, but are subtle and primarily rooted in functionality, and therefore the casual observer may not readily perceive the differences.

From a safety perspective, however, the distinctions are notable in that the controlled descent approach does not accommodate the very versatility that is

FIGURE 1-4
A typical controlled descent system.

foundational to safety in rope access. Specifically, the controlled-descent type system is not friendly to ascending, nor is it conducive to making transitions such as knot passes, rope-to-rope transfers, rebelays, direction changes, and so on, and it does not facilitate safety and simplicity for coworker-assisted rescue.

When workers fall into their backup system while using controlled descent methods, they are left suspended from whichever conventional fall protection attachment they are using. While in some cases this may be a sternal attachment, the dorsal attachment is arguably a more common point of attachment for conventional fall protection. A worker suspended from this point after a fall, as shown in Figure 1-6 is left without any means of escape. While it is possible to extricate oneself or another from a fall arrest system (McCurley, 2013) the fact remains that the methods for this must be learned, and generally require special training and equipment.

Another essential safety consideration is that there is no specified universal training requirement or certification format for controlled descent; typically, these methods are simply employed by people who are familiar with the techniques from using them recreationally. Worker training for conventional fall arrest, which typically consists of only up to around 10 hours of broad instruction that does not necessarily include practice for rescuing oneself or workmates, also does not normally address these methods. Contrast this with rope access training, which

FIGURE 1-5
(a) fall arrest harness (b) rope access harness. Courtesy of EDELWEISS SAS.

typically involves at least 40 hours of specific instruction in rope access methods and techniques, including rescue methods, just to authorize a technician to work under the specific direction of a more competent technician.

Rope Access Contrasted with Rope Descent Techniques

In recent years, the term "rope descent techniques" (RDS) has been tossed about in lieu of the term "controlled descent" but there appears to be little or no practical difference between the systems generally described by these two terms.

Rope Access Contrasted with Bosun's Chairs

The concept of the bosun's chair originated at sea, with the proper term being "Boatswain's Chair". A boatswain is the seafaring moniker for a warrant officer

FIGURE 1-6
Worker suspended in conventional fall arrest. Vertical Rescue Solutions by PMI.

who is responsible for operations on deck. In the case of sailing ships, part of a boatswain's responsibilities consists of maintenance and rigging of the sails. The boatswain's chair (bosun's chair), then, consisted of a wooden plank rigged with stout lines to form a seat, which was then suspended from the mast and raised and lowered by means of an integrated block and tackle, as shown in Figure 1-7. This was used to haul a sailor up the mast on a halyard to do repair work or inspect the rigging.

Humans, being the ingenious creatures that they are, adapted the concept of the seafaring bosun's chair to other types of work at height, particularly in coastal regions where sailors sought ways of plying their skills and earning a living ashore. The bosun's chair was quickly perceived by window cleaners, chimney sweeps, and even construction workers to be more secure, and at times more versatile, than a ladder, especially where a worker needed to spend extended time at height.

Today's version of a bosun's chair system has evolved to include greater emphasis on comfort, weight, and strength, and may also feature a body support strap or straps to help prevent the user from slipping out, as shown in Figure 1-8. These systems are typically used and regulated as suspended scaffolds, and should not be confused with rope access techniques.

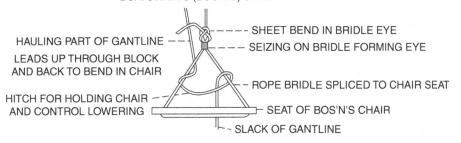

FIGURE 1-7
Classic bosun's chair. National Archives and Naval Historical Ships Association.

FIGURE 1-8
Modern bosun's chair. Courtesy of Tractel.

FIGURE 1-9
Rope access technician with comfort seat. Vertical Rescue Solutions by Pigeon Mountain Industries, Inc.

To be clear, some rope access technicians use a comfort seat that may, to the uninitiated, be confused with the concept of a bosun's chair. However, there are some distinct differences between the two. The rope access system incorporates dual, redundant, interchangeable rope systems, whereas the bosun's chair provides only the access component. The means of backup in a bosun's chair is typically some sort of vertical lifeline system, or self-retracting lanyard. In a rope access system, the worker's harness is directly connected to the suspension system, so that they cannot inadvertently "fall out" or become detached, while in a bosun's chair system it is typically the chair (rather than the harness) that is rigged to function as a single-point suspended scaffold and serve as the primary point of suspension for a worker. The comfort seat used by rope access technicians, as shown in Figure 1-9, is not a part of the life safety system.

Rope Access Backup Contrasted with Other Backup Systems

The compatibility of any backup system's ability to integrate with the primary system is essential. Rope access is particularly unique in that it provides BOTH the means of access and the means of height safety. In other systems, the means of access tends to be distinctly unique and separate from the means of safety. For example, a worker climbing a fixed ladder may be protected by a vertical lifeline; an inspector on a building façade may be protected by a fall arrest lanyard; a machine operator atop a hopper may be protected by a guardrail; an individual ascending stairs may be protected by a handrail; a construction worker may be protected by a safety net; and so on.

Building on this concept, then, a person who is rappelling but using a self-retracting lanyard (SRL) for backup is not using a rope access system; in this case, the descent line might be said to be a positioning system, and the SRL would simply be fall arrest. The fact that the two lines are not interchangeable, that the practitioner is not able to pass knots or deviations using these components, and that the duality required for coworker-assisted rescue is missing, disqualifies the system as a true "Rope Access System."

Where Modern Rope Access Has Landed – The Bus You Take to Work

Rope access is not, by itself, a viable trade. The value of a rope access technician being able to access, work at, and egress a worksite is only useful inasmuch as the work they are able to accomplish at that work location. Glazers use rope access to reach locations where they repair and install glass; engineers use rope access to reach locations where they inspect buildings, bridges, and other structures; wind turbine technicians use rope access to reach locations where they inspect and repair wind turbines; entertainment riggers use rope access to reach locations where they rig and perform stunts and set-work; oil rig workers use rope access to reach locations where they maintain offshore facilities; window cleaners use rope access to reach locations where they clean windows on very tall buildings; rock scalers use rope access to reach locations where they mitigate roadway hazards; and the list goes on.

The act of rope access, then, is useful in that it permits a tradesman to get to a given location where they can apply their trade. With this in mind, rope access is best thought of not as a trade, but simply as a means of access. In essence, rope access can be said to be *"like the bus you take to work."* The term does not describe the work itself, only the means of accessing the worksite.

1-4 COMPATIBILITY AND WORK SAFETY

The compatibility of rope access with other methods of protection must be given due consideration in planning any work. Different countries and jurisdictions may have varying degrees of regulation that can be applied to rope access but, as a general guide, International Standard ISO 22846 provides an excellent framework for the specification and execution of rope access as a work practice. This practice can be used in harmony with other means of access and safety protection, with the worker transitioning between using rope access methods, positioning, fall arrest, and others.

To ensure that a rope access system operates correctly, the employer must adequately plan and manage the work, ensure competency and supervision of workers, provide appropriate equipment, and effectively organize working methods so that the worker is adequately protected at all times.

Rope access methods are often combined with conventional fall protection methods (restraint, positioning, arrest) to accomplish engineering inspections, maintenance work, installations, and other tasks in an efficient and comprehensive manner. The initial training for an entry-level rope access technician may be in the neighborhood of 40 hours, while advanced certification requires extensive experience and verification of greater skill levels. Such extensive training and skills help ensure the ability of the workers to effectively interweave their use of rope access systems with other conventional means of protection.

Fall Restraint

Rope access technicians may at sometimes use Fall Restraint methods once they reach their place of work. A Fall Restraint system will typically incorporate a safety belt or harness, a lifeline or lanyard, and an anchor, which are used together to prevent the worker from reaching an edge where fall potential might exist.

Restraint systems are most often used in an environment where temporary work is being performed, and where the exposure, equipment, and training necessary for a full-blown fall arrest system are simply not justified. A key requirement of a restraint system is that the worker's center of gravity does not reach fall potential.

While restraint systems are often used by workers with considerably less training than rope access technicians are likely to possess, this type of system may also sometimes be useful to trained rope access technicians who have reached a work location via rope access methods, and simply want to remove themselves from any fall potential. In this case, the technicians may connect themselves to the restraint system and then (as long as all fall potential is mitigated) remove themselves from their working line and backup line, or they may elect to leave their backup safety (and possibly even working line) connected to them while simultaneously engaging the restraint system.

Fall Arrest

Fall arrest is a conventional approach to protecting a worker who is at a place of work where fall potential exists. The term Fall arrest does not describe a method of access, but refers only to the system used for secondary protection. These methods are typically used when the worker's feet comprise the means of access, such as when they are walking a beam, climbing a ladder, or standing on a platform.

FIGURE 1-10
Fall arrest. Courtesy of Reliance Industries, LLC.

Use of personal fall arrest systems should only be considered where better protection is not reasonably available. Normally a fall arrest system, illustrated in Figure 1-10, will consist of an anchorage, lanyard, force absorber, connectors, and a full body harness. Additional components, such as a rope, a fall arrester, and other equipment, may be incorporated into some systems.

While personal fall arrest systems are perhaps the most common means of protecting workers at height, they are arguably not the best choice for protection. In practice, users of these systems are potentially exposed to significant free fall, high arrest forces, and have the potential for striking obstructions before or during the arrest phase of the fall. Depending on the jurisdiction, the maximum allowable arresting force on the worker ranges from 1800 lbf (8 kN) to 900 lbf (4 kN).

Fall arrest systems come in many forms, ranging from a simple force-absorbing lanyard to vertical lifelines to self-retracting lanyards to horizontal lifelines to ladder climbs, and more. Due to the complexity of calculating potential loads, forces, strengths, and consequences of a fall, the design and approval of such systems often necessarily require extensive knowledge and experience of a qualified person. Authorized users of conventional fall arrest systems are not necessarily required to possess extensive training or experience; however, the systems specified by qualified persons generally employ equipment and methods that are neither versatile nor interchangeable.

Most rope access technicians will strive to avoid using conventional fall arrest systems, and instead favor the limited fall potential and greater comfort, security, and versatility of a rope access backup system.

Fall Containment Systems

It is rare for rope access systems to be used in conjunction with a fall containment system. The term containment system refers to passive fall arrest techniques such as safety nets that actually permit a fall to occur, but are designed to catch the worker before any catastrophic impact. Again, this term does not describe a means of access, but only the means of secondary protection. Most commonly found in bridge work and steel erection projects, where other methods of protection are not possible, safety nets may be rigged as much as 30 ft beneath the working surface.

Because rope access is primarily used as a robust alternative to infeasibility, nowadays it is not unusual to find rope access solutions in places where fall containment systems may have previously been used.

Suspended Scaffolds

The term suspended scaffold refers to any of a number of types of platforms suspended by ropes, or other nonrigid means, from an overhead structure. The most common type of suspended scaffold features a two-point adjustable suspension system and is commonly referred to as a Swing Stage, an example of which is shown in Figure 1-11. It is common to combine rope access methods with the use of this or other types of suspended scaffold during building maintenance work, bridge inspection and maintenance, wind turbine work, and in other environments.

Some means of backup safety must be used when working on a suspended scaffold. Typically, suspended scaffold workers will use conventional fall arrest for this purpose. When using rope access in conjunction with suspended scaffolds, it may be appropriate to temporarily use the rope access backup system as backup safety while on the scaffold, as long as the limitations of the rope access backup system are not exceeded.

In any case, the worker's backup system should be secured to an appropriate anchorage that is separate from the scaffold system.

Aerial Platforms and Man Baskets

Another solution that is frequently implemented where other means of access and protection are infeasible are aerial platforms and man baskets, such as that shown

FIGURE 1-11
Workers using two-point suspended scaffold.

FIGURE 1-12
Workers using aerial platform. Courtesy of Reliance Industries, LLC.

in Figure 1-12. When there is no surface from which a worker can perform a task, aerial baskets, bucket trucks, scissor lifts, snoopers, and cherry pickers may be used to temporarily raise a worker to an elevated position to accomplish that task.

Note that aerial platforms and man baskets are a means of access; secondary protection is generally provided to the worker by means of a handrail or conventional fall arrest system.

It is common to find rope access methods employed harmoniously with mechanical lifts such as these. Especially in complex environments where an extensive duration of work is combined with the need to reach elusive nooks and crannies, aerial lifts can provide an excellent platform from which to set anchors and serve as a work-base for rope access technicians to further access more remote areas. One example of this would be to perform work or inspection on the underside of bridges.

1-5 SPECIAL TECHNIQUES

There are some special techniques in which rope access technicians are trained but which do not meet the strictest definition of rope access; there being the presence of a primary line for ascending and descending, and there being an equal and interchangeable line for backup safety. The most notable of these are lead climbing and aid climbing, so by way of introduction those concepts will be outlined here. These and other advanced techniques will also be covered in further detail in Chapter 11.

Aid Climbing

In some cases, a work location may not be conducive to installation of a fixed rope; in such cases, aid climbing may be a viable solution. This method, illustrated in Figure 1-13 and further described in Chapter 11, involves the practice of progressing from one anchor to another by the continuous placement of slings and footloops, into which the technician clips lanyards for protection and places their feet into for progression.

A properly trained rope access technician will never have less than two points of attachment at any given time, and potential for a fall is mitigated to less than 2 ft.

FIGURE 1-13
Aid climbing. Courtesy of Abseilon USA, www.abseilon.com.

Lead Climbing

Where a worker must access a structure from below or from a parallel position, without the advantage of a preset anchorage or lifeline from above for protection, lead climbing may be an appropriate technique. Lead climbing involves the placement of anchorage protection and continuous attachment to that protection while climbing a structure. Lead climbing techniques may also be used to protect leading edge work.

Of all the methods used by rope access technicians, this one offers the highest fall potential, so it is only used when absolutely necessary. Specific techniques for lead climbing are discussed in Chapter 11.

1-6 PRACTICAL APPLICATION OF ROPE ACCESS

Responsible employers around the world use rope access methods and techniques that have been implemented by certified and competent technicians; these methods have helped to ensure 100% protection in some of the most difficult to reach locations and in some of the harshest environments, such as the one shown in Figure 1-14.

Rope access has been used to inspect the dome of the US Capitol Building in Washington DC, to pressure-wash the outside of the 605 foot Space Needle in Seattle, Washington, and to clean the glass on the under-side of the Grand Canyon Skywalk, some 4000 ft above the canyon floor. Rope access is commonly used in hot, cold, windy, and even icy conditions, ranging from chilly places as far north as the Arctic to the sweltering sites on the Equator, with equal safety and security.

The cost of implementing a rope access program in a given organization will vary depending on a number of factors. Basic training and personal equipment for one technician, including certification fees, can be estimated at approximately $3500. A minimum of two technicians within an organization should be trained as a starting point – although a team of 6–8 is really recommended for an organization

FIGURE 1-14
Rope access system accommodates safety in hard-to-reach places. Courtesy of Abseilon USA, www.abseilon.com.

that is doing any significant amount of work. A basic complement of rigging gear, including ropes, connectors, and anchorage equipment, can run in the $5000 range, bringing the initial personnel and equipment investment to a minimum of around $15,000 for a team of two.

Because supervision of rope access jobsites is a specialized skill, it is common for organizations to outsource jobsite supervision at first, until staff technicians are sufficiently experienced and certified to perform this task. However, it is safe to say that a basic, simple rope access program can be put into effect for around $15,000. Provision for retraining and equipment replacement should be scheduled on at least a 3-year rotating basis, although specific requirements will vary.

1-7 SUMMARY

The application of rope access methods is most appropriately engaged as a complete system of work in which the following are present and carry equal importance:

- adherence to an accepted rope access plan;
- employment of only competent and certified technicians; and
- exclusive use of equipment that is purpose-designed for rope access.

This text is intended for use by all persons concerned with the use of rope access, including technicians, managers, safety directors, supervisors, employers, trainers, clients, and regulatory personnel.

CHAPTER 2

Rope Access and the Comprehensive Managed Fall Protection Plan

By the end of this chapter, you should expect to understand:

- How the term "employer" is intended to be understood in this text
- The importance of a comprehensive fall protection program
- The principles behind a successful fall safety program
- The duties of the Program Manager
- How Authorized, Competent, and Qualified Persons differ
- The several documents within the Comprehensive Managed Fall Protection Program
- The hierarchy of Protection against Falls from Height
- The several harness-based methods
- The contents of a Rope Access Work Plan
- The essential components of a rope access work system.

2-1 PROTECTING WORKERS AT HEIGHT

Few would disagree that the employer has a responsibility to provide a safe and healthful workplace. This text will not attempt to wade through the complex matter of division of legal responsibilities of owners, employers, contractors, or other jurisdictional layers within the workplace. This is a matter that legal authorities have battled over for decades, and will undoubtedly continue to do so as they seek to protect the assets of various entities. The finer points of that subject are beyond the scope of this book.

This discussion of employer's responsibilities will make numerous references to "employer" within the context of their being the authority having a jurisdiction over the workplace. This is not to suggest that there are no other entity(ies) who might also have authority or responsibility, but it is simply to capture the essence of the generally defined and established responsibilities of a business entity to protect its

---

Professional Rope Access: A Guide To Working Safely at Height, First Edition. Loui McCurley.
© 2016 John Wiley & Sons, Inc. Published 2016 by John Wiley & Sons, Inc.

employees. Where the term "employer" is read in this chapter – indeed, throughout this text – in some cases it may be appropriate to substitute with the term(s) "owner," "operator," "contractor," "subcontractor," "manager," or other term, depending on jurisdictional requirements and authority.

That said, it should also be understood that there is a corresponding responsibility upon each and every individual worker to take a certain amount of responsibility for their own personal safety.

A successful safety program requires effective planning, training, and management. While there is no one magic tool that applies to all circumstances, there are some general principles that are shared by those who work safely at height. This chapter will explore those principles in relation to the overall comprehensive managed fall protection plan.

2-2 COMPREHENSIVE MANAGED FALL PROTECTION

A notable resource that is particularly valuable for safety managers who oversee workers at height is the ANSI Z359.2 standard for a Comprehensive Managed Fall Protection Program. The term comprehensive, simply meaning "complete," serves as a reminder that the fall safety plan must encompass all potentially affected individuals, and as a reminder of the importance of thorough preparation. This standard, designed with employers in mind, asserts that the most effective method of protecting workers at height is to develop and implement a *Comprehensive Managed Fall Protection Program*. It provides an excellent guideline from which to develop one's own fall protection plan. It walks the user step-by-step through the development of such a plan, and offers insight and direction at every level. While not specific to Rope Access, it is an excellent precursor to what the rope access community refers to as a Rope Access Work Plan (sometimes also called a Rope Access Permit.)

The Comprehensive Managed Fall Protection Program document should be written to address all of the topics outlined here, as well as any additional topics that may be appropriate to a given employer's specific circumstances. While not all regulatory authorities require that a document be written, having a written plan makes a plan arguably more conducive to review, train, and adhere. In addition, a written document is much easier for an employer to stand upon in the event that it does come under regulatory or legal review.

Policy Statement

According to ANSI Z359.2, the first step in developing a Comprehensive Managed Fall Protection Program is to adopt a policy statement that provides general goals and guidance related to the employer's intent. The statement should emphasize the management's commitment to providing a safe workplace for employees who may be exposed to fall hazards, and the tone of the statement will effectively set the stage for the program as a whole.

> **Sample policy statement from ANSI Z359.2**
>
> The (employer's) safety policy is to take every reasonable precaution to protect the health and safety of employees. Implicit in the safety policy is the requirement that employees shall use effective fall protection systems when working in any situation that presents a foreseeable exposure to a fall hazard.

Chapter 15 offers a process that employers, site owners and safety managers can use in developing a policy statement for their organization.

Staff Responsibilities

Ensuring clear identification and communication of individual responsibilities will help to prevent duplication of effort while also ensuring that identified needs are addressed.

ANSI Z359.2 identifies the employer as being ultimately responsible for providing the fall protection plan, as well as the resources required to support that plan. A complete fall protection plan, according to Z359.2, includes a hazard assessment, equipment descriptions, training records, rescue provision, and an adequate structure for management/supervision.

In regulatory-speak, the terms Qualified, Competent, and Authorized are used to describe the levels of worker responsibility and respective permissions on the part of the employer. As a general rule, the term "Authorized Person" refers to an individual who is approved or assigned by the employer to perform a specific type of duty or duties or to be at a specific location or locations at the jobsite. The term "Competent Person" is used in reference to one who is capable of identifying the existing and predictable hazards in the surroundings or working conditions that are unsanitary, hazardous, or dangerous to employees, and who has the authorization to take prompt and corrective measures to eliminate them. The term Qualified Person generally means *one who, by possession of a recognized degree, certificate, or professional standing, and/or who by extensive knowledge, training and experience, has successfully demonstrated the ability to solve or resolve problems relating to the subject matter, the work, or the project.*

It is imperative within the context of any safety program for the employer to clearly denote which employees are given which designation. Just because an employee has received training to a given level – for example, as a Competent Person, this does not make them a Competent Person. It is only the employer who can give an employee such a designation, within the context of a particular form of work. Speaking specifically of fall protection rope access programs, the employer should clearly state for each employee whether they are working as an Authorized, Competent, or Qualified Person. It is also the responsibility of the employer to identify a Program Administrator to oversee the fall protection program, and to ensure that competent and authorized persons are adequately trained and supervised to accomplish their work safely.

The designated Program Administrator is generally responsible for overseeing the overall fall protection program, including preparation, implementation, and review. This is not to say that the Program Administrator performs all of these duties personally, only that this designated person is to ensure that these necessary tasks are properly assigned and performed. This role need not be an exclusive designation. A Program Administrator may also serve within the scope of the Comprehensive Managed Fall Protection Program as a Qualified or Competent Person, Trainer, and/or Rescuer.

Although the Program Manager need not be directly responsible for systems-specification, anchorage selection, or rigging, it is imperative that the Program Administrator possesses a thorough knowledge and understanding of relevant fall protection regulations, equipment, and methods. Where a Program Administrator will oversee a Rope Access Program, it is essential that the Program Administrator be thoroughly familiar with rope access, including holding (or having previously held) appropriate credentials specifically as a rope access technician.

Given the unique integration of equipment, systems, and management required in a rope access system, a first-hand understanding of this interconnectedness is crucial.

The Program Administrator will assign duties and responsibilities to affected personnel, to ensure that procedures have been established for protection and rescue as necessary, oversee the provision of adequate resources to achieve that plan, review the program periodically, and participate in the investigation and corrective actions relative to incidents and near misses.

The Program Administrator may also function as a Qualified Person, and/or may secure the services of others from within or outside the organization to provide additional input in this capacity. A Qualified Person is simply a technical resource that supports the fall protection program in some capacity of expertise such as system design, structural analysis, calculation of forces, certification of components, compliance, and so on. Again, a thorough and specific understanding of all systems for which they are responsible including regulation, industry best practice, equipment, systems, methods, and physical sciences are essential. Again, in cases where rope access is used, the Qualified Person(s) must necessarily be specifically familiar with rope access, ideally being (or having been) a properly credentialed and experienced technician.

It is also possible for the Qualified Person to serve simultaneously or successively in the capacity of a Competent Fall Protection Person. It is most appropriate that the Competent Fall Protection Person be assigned the immediate supervision of work at height, including implementation, supervision, and monitoring of the managed fall protection or rope access program. This role is typically assigned to an individual who is very knowledgeable in the specific area(s) in which they are deemed "competent," and has sufficient experience to make decisions based on judgment, and who holds a position of some authority so that he may stop work immediately if unsafe conditions warrant. Note that a Competent Fall Protection Person may or may not also be deemed a Competent Rope Access Technician, and/or Competent Rescuer. Competent Persons must be adequately trained specifically in the area of specialty before being deemed "competent" by the employer.

The Competent Person should work under the direction of the Program Administrator to prepare written procedures, implement their use with all affected persons, and control exposure of authorized persons on a day-to-day basis. These responsibilities also include the duty of ensuring that prompt rescue of authorized persons will be available in the event of a fall. In this case, the definition of "prompt" may vary depending on specific circumstances such as potential hazards and the condition of the fallen subject. For more information on post-fall rescue from fall protection, see *Falls from Height: A Guide to Rescue Planning*.[1] Other duties that are often assigned to a designated Competent Person include equipment selection, inspection and retirement, and investigation of incidents.

Not everyone who works at height needs to be designated by the employer as a Competent Fall Protection Person. Most individuals who are assigned or permitted to work at height are generally referred to simply as an Authorized Fall Protection Person – or as an Authorized Rope Access Technician, or an Authorized Rescuer, depending on the specificity of their training and experience.

All Authorized Persons must have a reasonable understanding of the procedures and instructions under which they are expected to work, but should be working at all times under the direction and supervision of a Competent Person. Although the Authorized Person may not have as extensive experience or knowledge as does the Competent Person, they should be diligent enough to ensure their own safety. In doing so, they should adhere to the established practices and procedures set forth by the employer, use PPE and other equipment properly, and

---

[1] *Falls from Height: A Guide to Rescue Planning*, (Wiley, 2013) 376 pages. Author: Loui McCurley ISBN: 978-1-118-09480-8

should also use good judgment in immediately reporting any unsafe or hazardous conditions to a Competent Person.

At every level, potential trainers must be knowledgeable and experienced at a level higher than that of the students they will be expected to teach, and must have a thorough understanding of the environment and hazards applicable to the industry for which they are training personnel.

Fall Hazard Survey

Employers should ensure that a fall hazard survey is performed at any and all locations where an Authorized Person may be exposed to a hazard, and that a report is prepared to identify these hazards as well as ways and means by which the hazards may be eliminated or controlled. Tools for developing the fall hazard survey, which should receive input from affected employees and managers (Figure 2-1), may be found in Chapter 19.

While specific requirements may vary depending on jurisdiction, it is generally accepted that workers must be protected when they are working above 6 ft. In fact, according to U.S. federal regulations, fall-protection must be provided if an employee can free fall 4 ft or more onto a lower level when engaged in general industry or 6 ft or more onto a lower level when engaged in construction. This includes (but is not limited to) situations where an employee:

- is on a surface that has an unprotected side or edge;
- is constructing a leading edge;
- may fall through a hole in the walking/working surface;
- is working on the face of formwork or reinforcing steel;
- is on a ramp, runway, or other exposed walkway;
- is working at the edge of an excavation, well, pit, or shaft;
- is working above dangerous equipment;

FIGURE 2-1 Get input from workers for fall hazard survey. Courtesy of Abseilon USA, www.abseilon.com.

- is performing overhand bricklaying and related work;
- is performing roofing work on low-sloped roofs;
- is working on steep roofs;
- is engaging in precast concrete erection (with certain exceptions); and
- is engaged in residential construction (with certain exceptions).

It is the responsibility of the employers to be aware which rules and regulations apply to their workplace. Where there are multiple potentially applicable rules that appear to be contradictory, it is always best to take the most conservative approach. The fall hazard survey should be performed by someone who is familiar with the requirements of the country and jurisdiction in which the work is being performed.

In most cases, a Competent Fall Protection Person or Competent Rope Access Technician should prepare the Fall Hazard Survey Report. A well-developed Fall Hazard Survey Report provides essential information that the Program Administrator and other affected individuals will use throughout the course of the work. The report should contain pertinent information regarding types of fall hazards, detailed information as to the specific configuration of the hazard (including drawings or photos), analysis of how severe the exposure potential might be, information regarding the frequency and duration of the work, recommended corrective solution(s), and rescue procedures in the event that a fall does occur.

Extraordinary factors such as environmental conditions, obstructions, work tools, area contaminants, other workers in the area, and other potential hazards should also be considered. In addition, the report must address installation, use, inspection, maintenance, and de-rigging of any fall protection systems intended to be used. Records of previously known incidents at the location addressed by the report, as well as at similar locations, can be a useful tool when performing a fall hazard survey.

The Fall Hazard Survey Report is a living document, and should be revised or re-written anytime a substantial change is made to the task, process, structure, equipment, or pertinent legislation.

2-3 HIERARCHY OF FALL PROTECTION

Wherever a potential fall hazard is identified, methods of protection should be selected to provide the maximum possible safety for the worker. To this end, the "hierarchy of protection," illustrated in Figure 2-2 as an inverted pyramid, offers some insight on how to approach the question as to which type of protection might best fit your application.

In this illustration, fall protection methods are presented in an order that is presumed to preferentially provide the greatest safety. At the top of the inverted pyramid lies the approach that many believe to be most safe, with the methods that still expose the worker(s) to greater risk nearer the base.

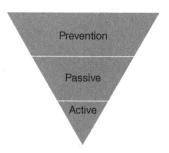

FIGURE 2-2
Hierarchy of fall protection.

There have been numerous approaches to this hierarchy over the years, but the most recent example comes from discussions underway within the ANSI Z359 standards-writing body as this text goes to print.

Foundationally, this diagram illustrates the idea that the "safest" means of fall protection is to simply avoid the hazard. Titled in this illustration as "Prevention," the idea refers to any method that eliminates that hazard, or at least prevents exposure to it. The terms "Elimination" and "Substitution" have also been used to denote this approach to fall protection. There are any number of means by which prevention may be achieved, but regardless of the approach the result should be the same: the potential hazard is removed.

If the hazard cannot be removed, the next means of protection that should be considered is passive protection – that is, any means of preventing a fall by separating the Authorized Person from the fall hazard. Essential to the success of passive protection is that it is not reliant on the actions of the worker. A minimal amount of training should be required for its use. In most cases, this would involve a physical barrier, such as guardrails or covers, to separate the worker from the fall hazard. Guidance for passive protection may be found in ANSI/ASSE A1264.1 (ANSI). The problem with passive protection, and the untrained persons who may be in the vicinity of the work, is that workers are notorious for disregarding, or even disabling these would-be protective measures. Examples abound of workers losing their lives as a result of stepping over a guardrail, crawling through a barrier, or removing a cover "just for a moment" to accomplish a task that is just beyond reach.

The final category of protection is commonly referred to as active protection. Active protection denotes any means that involves the use of harness-based systems to protect the worker. This would include such methods as restraint, rope access, positioning, or fall arrest. The concern that is most often cited with harness-based protection is that these methods rely upon worker training and equipment function for maximum safety. On the flip side of this argument, however, is the fact that better-trained workers have a wider array of tools and knowledge at their disposal to maintain their own safety.

Distilling height safety down to these three categories helps remind the safety manager that exposing a worker to less risk is always the easiest solution.

Types of Active Protection – Harness-Based Solutions

Harness-based protection is a little more difficult to reduce to a hierarchical summary. Some people subscribe to the utopian idea that work involving equipments that could be misused should be avoided at all costs, or at the very least the equipment should be engineered so as to remove any potential for human error. Others argue that our lives are enhanced through the use of equipment that is operated by humans (automobiles, tools, equipment, etc.) and that the best solution is not to eliminate the tools but to improve the knowledge and skill of those who use them.

The best solution for any given circumstance depends on several factors, including frequency and duration of exposure, ability to effectively train and equip workers, and the type of work that is to be performed.

If harness-based methods are deemed appropriate, rope access is an excellent choice that offers unsurpassed protection to permit a worker to be suspended securely with a secondary line as backup for security. Specialized training and techniques are required for this approach, making it the most training-intensive option of all harness-based alternatives – and for the same reason it is also the most versatile. Rope access technicians are better equipped and trained to protect themselves when they find themselves in circumstances that do not fit textbook

FIGURE 2-3
A positioning system, used with fall arrest for safety. Courtesy of Pigeon Mountain Industries, Inc.

fall protection scenarios. This results in a clear advantage for workers to be able to access the hardest-to-reach places with minimal impact and maximum safety.

Fall restraint, an approach that involves tethering the authorized person to a suitable anchorage using a lanyard short enough to prevent the person from being exposed to the fall hazard, is often cited as the next most preferred method of harness-based safety. This method requires less training than rope access, but significantly limits the places that the worker can reach. One hazard of this method is that users may be tempted to momentarily disconnect their system as they attempt to reach a "just out of reach" location.

The concept of restraint should not be confused with that of positioning, which refers to a system that holds the workers in place while keeping their hands free to work. A positioning system is activated by the action of a worker leaning into it, as shown in Figure 2-3. A positioning system is not the same as restraint, nor is it designed for fall arrest purposes.

If other methods are not viable options, a worker may be protected using a conventional fall arrest system. The term "fall arrest" broadly encompasses a variety of systems designed to stop an authorized person after a fall has begun. Fall arrest systems range from fall arrest lanyards with force absorbers to self-retracting lanyards to vertical or horizontal lifelines.

Choosing a System of Protection

In selecting methods of fall protection for a work-at-height program, consideration should be given to training and qualifications required for each different approach, and the impact this may have on its effectiveness. Where infrequent exposure occurs, simple systems like lifts and guardrails may be a good solution because they require minimal training. For workplaces where more frequent exposure occurs, consideration should be given to the idea that more thoroughly trained workers may be better prepared to accept responsibility for their own safety.

Of course, more than one means of control may be used simultaneously or consecutively for a given situation, if the employer deems this to be appropriate. In any case, the fall protection plan should be specific to the workplace and task where it will be applied and should be designed to provide for 100% continuous fall protection.

Where active fall arrest systems are selected as the means of protection, the employer should follow accepted industry guidelines for protection and provision for rescue. Active fall protection systems require complete analysis and consideration of anchorages, equipment, clearance requirements, access/egress to the system, installation/derigging, and rescue planning.

Before adopting rope access as a method for a given situation, a risk assessment should be performed to consider all alternatives available as well as assess their respective potential advantages and hazards. In particular, the Program Manager must consider the consequences of any method that is chosen and whether the method is conducive to the particular work and also ensure that the tools necessary for the job are at hand.

In addition, consequent to the others above or below, workers or bystanders must also be considered. In some cases, provisions may need to be made to blockade hazard zones or cordon off areas of access for reasons of security of the work being done as well as those of the others in the vicinity.

In all cases, duration and levels of exposure, and the influence on risk, should be considered.

2-4 ROPE ACCESS WORK PLAN

Once the fall hazard survey has been completed and it has been determined that rope access is warranted, the employer should move on to preparing a Rope Access Work Plan, also sometimes referred to as an Access Permit. The concept of an Access Permit, or Rope Access Plan, is not unlike the permit process used in the United States for confined space entry. This concept places a direct and specific responsibility on the employer's Competent Person to follow certain steps before issuing a permit to an Authorized Person to enter the space.

Section 3 of this book is designed with respect to planning, and Chapter 17 provides a guide for developing a work plan for rope access. That section is intended to be used as a step-by-step reference, with pen in hand, once the safety manager is fully conversant with the idea of rope access, and how it fits into the overall managed fall protection plan. It is highly recommended that the reader completes going through Sections 1 and 2 of this text to gain a greater understanding of rope access prior to attempting to use Section 3 to write a rope access plan.

Within the context of rope access, the access permit should address the following safety objectives, to include (but not be limited to) the following:

- List the rope access methods to be used for the proposed work.
- List the members of the work team by name and identify their duties.
- List the rope access equipment to be used for the work to be performed.
- List the hazards associated with the work to be performed (JHA).
- List appropriate personal protective equipment (PPE) to be used.
- List provisions for providing security to the anchor.

- List public safety provisions.
- List the rescue service and the means to summon the rescue service.

Additional requirements may need to be considered depending on the work situation and the work task being performed.

The rope access plan, when applicable, is really just one component of the comprehensive managed fall protection plan. As such, it is most appropriately prepared by a Competent Person under the direction of the Program Administrator.

System Requirements

The greater majority of this text deals with specific methods and techniques used for rope access. These methods include techniques for descending, ascending, and moving horizontally across multiple sets of rigged ropes while in suspension or tension. Special techniques are also used for protecting oneself while using a structure for support.

Regardless of the particular technique used, the principle of double protection is fundamentally imperative to the concept of rope access. In effect, if any one component within the system should fail, the worker would be kept safe by the other system.

In addition, integral to the concept of rope access is the matter of the secondary system being fully compatible and interchangeable with the primary system. This is not to say that it is not appropriate for a conventional fall arrest system to be used as a backup for a person descending the rope. However, such a system would not be considered rope access. As explained in Chapter 1, it would simply be considered to be a descending system wherein the worker is protected by conventional fall arrest (Figure 2-4).

Before any work is undertaken, the specific roles and responsibilities of each team member should be clearly communicated. One team member, usually one with specific management experience and a thorough understanding of the work to be performed, should be designated for being responsible for supervision. On a worksite with more than one distinct or isolated working area, the employer should ensure that adequate supervision is assigned to each of those areas.

Potential falls should be protected in a manner 'that prevents the worker from taking an impact on the ground or from injury by impact with structures or

FIGURE 2-4
Conventional fall protection used with rope descent is NOT the same as rope access. Courtesy of Abseilon USA, www.abseilon.com.

obstructions. It is also generally accepted that falls be mitigated to impart no more than 6 kN of force on the technician, and that uncontrolled swings be prevented. Avoiding slack in the safety line helps to achieve these goals.

For the safety of the worker, systems should be rigged to minimize potential fall distance and prevent entanglement or hazardous contact with machinery, power lines, or other obstructions. In addition, measures should be taken to ensure that the worker cannot involuntarily descend off the end of the rope.

The rope access technicians should always connect to and disconnect from their system at a location where no fall hazard potential exists, or should use alternative means of fall protection while connecting/disconnecting, as needed. It is generally good practice for technicians to make their connection to the backup system even before attaching their access system to the rope, and at the end of the work to remove their backup connection only after connection to the access system has been disengaged.

A rope access technician must wear an appropriate full body harness with at least waist and sternal connecting points that facilitate connection to both the access system and the backup system at all times. Multiple access or safety lines may be – and often are – connected to the same D-ring or attachment point on the harness. However, in most cases only one system at a time will be loaded. The worker's harness should be connected via appropriate lanyards directly to both the access system and the backup system, even if a workseat is being used.

Appropriate equipment for rope access is discussed in Chapters 4 and 5. Suffice to say here that all equipment should be used within the scope of its designed purpose and intent, and should be capable of withstanding foreseeable forces with an adequate safety factor. Suitability of equipment may be driven in part by such factors as performance specifications, regulatory requirements, and type of access methods being used.

Modern ropes and equipment used for rope access are quite robust, highly reliable, and very strong. Nonetheless, appropriate precautions should be taken to protect ropes and other equipment from environmental hazards. Sharp or abrasive edges may be padded, or specially designed rope sleeves may be used for this purpose. Equipment protection should also be considered where falling debris could pose a hazard. Equipment should be thoroughly inspected by a Competent Person on a regular basis, especially prior to use by the worker intending to use it.

Safety in Rope Access Operations

Every worksite involves safety requirements that arise from the work being performed as well as from the methods of protection being used. It is the responsibility of the Program Administrator to ensure that all aspects of safety are addressed; however, different aspects of safety may be delegated as appropriate to different individuals.

When rope access systems are used, personnel should always work in teams of at least two, with one person designated as the site supervisor. The supervisor should be a designated Competent Person, with specific relevance to rope access, as the primary worksite safety is the responsibility of this person. The supervisor must be familiar with not only rope access, but also conventional methods, including the limitations and capabilities of all options (Figure 2-5).

Rope access operations often occur at remote worksites, or in locations that are not easily accessed by common emergency services or other conveniences. For this reason, a rope access team should be self-sufficient and capable of supporting their own needs throughout the course of the work.

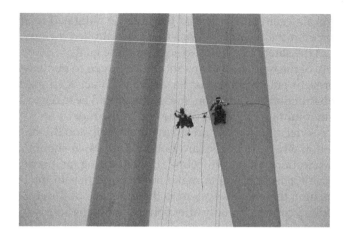

FIGURE 2-5
Rope access teams should consist of no fewer than two persons. Courtesy of Trask Bradbury.

Where necessary, personnel should be assigned and dedicated to perform specific safety functions such as traffic monitoring, controlling access to hazardous areas, monitoring anchorage areas, and other duties. Personnel performing these functions do not necessarily require rope access skills unless the position puts them in a place where those skills are needed.

Rescue preparedness is paramount in rope access. While site safety is primarily the responsibility of the supervisor, all workers should be trained in emergency escape ("self-rescue") from any potential circumstance in which he might be placed. In addition, the supervisor should ensure that provision is made for rescue of any worker who is working at height, in the event that the worker becomes incapacitated. Especially in cases where one technician is more skilled or experienced than the other, that technician must take care not to be put in a situation where there is not sufficient rescue capability in case of a fall. Note that it is not always necessary for a technician to enter the vertical environment to effect a rescue; in many cases ropes may be rigged in such a manner that retrieval may be performed from above or below, without the "rescuer" ever being suspended. These methods are further discussed in Chapter13.

All persons engaging in rope access should be properly credentialed by verification of skills by an appropriate body. Skills, knowledge, and aptitude should all be considered, as should familiarity with the specific environment in which work is being performed. Work team and personnel selection and considerations are addressed in detail as part of Chapter 3.

It is essential that the Rope Access Supervisor be capable of maintaining adequate communications with all team members to ensure safety. An effective communication plan should be established and all team members apprised prior to the start of operations. Communication systems and plans should be considered to ensure that they are compatible with the conditions of the worksite such as weather, noise, interference, as well as with other work teams. This may be accomplished by means of voice contact, hand signals, radio, or other means, as long as the entire work team has been apprised and fully understands the methods to be used. Consideration should also be given to communications with external resources such as other work teams on the site, emergency services personnel, and upper management.

Work Practices

Although it is the responsibility of the Rope Access Supervisor to ensure that these good practices for safety are maintained, all team members should take personal responsibility for safety at the worksite. A Rope Access Supervisor is typically a technician with an advanced level of certification and extensive knowledge, skills, and experience in the use of rope access in the workplace.

Prior to the start of a given work period, the Rope Access Supervisor should ensure that any equipment that may have been left rigged is manually inspected, paying particular attention to anchorages, ropes, and any point in the system where rope or equipment may have come into contact with abrasive surfaces or structural elements.

Likewise, at the start of each work period the supervisor should meet with the entire work team to collectively review the progress and ongoing plans for that day, as well as any risks that might affect safety and/or effectiveness on the job. All team members should understand the objectives for the day as well as work and safety procedures. Commonly known as a "tailgate meeting," documentation should be made as to who was present and what was discussed at this meeting.

Before work for a given day commences, life safety and other working equipment should undergo a pre-use inspection. Each technician should carefully inspect his own PPE, access equipment, connectors, and backup devices to ensure that they are in good working condition. Technicians should also be trained and encouraged to use "buddy check" procedures at the start of each day and throughout the course of the work. The term "buddy check" refers simply to the practice of inspecting a coworker's rigged harness and equipment to verify their readiness for work (Figure 2-6).

Likewise, if a second shift is taking over at the close of the work period, this transition should be made formally to ensure that the shift is adequately informed and briefed, and that they perform the necessary tailgate meetings, pre-use inspections, and buddy checks prior to the start of that shift. If the close of a workday involves shutting down of the work, the site should be cleaned and the equipment and work tools should be secured and stored carefully to prevent damage from the elements or tampering by passersby.

FIGURE 2-6
Technicians should be encouraged to use a Buddy Check system. Courtesy of Pigeon Mountain Industries, Inc.

2-5 SUMMARY

Rope access is but one potential method of access and protection that may be used within the context of a complete Managed Fall Protection Plan. This text provides sufficient information for an employer to use while developing a fully functional rope access program, and more. However, additional outside resources, such as experienced trainers, supervisory staff, and management consulting may be required if an employer does not have persons on staff who possess these skills.

In no case should this text be considered a replacement or substitute for hands-on training or experience.

CHAPTER 3
Personnel Selection and Training

By the end of this chapter you should expect to understand:

- The essential requirements for rope access personnel
- The difference between rope access and a skilled trade
- How to evaluate aptitude for rope access
- The makeup of a rope access team
- Technician skills and responsibilities
- The responsibilities of the Rope Access Program Manager
- The responsibilities of the Rope Access Supervisor
- The essential requirements for training
- The importance of skills verification
- How to select a Certification Body.

3-1 INTRODUCTION

As with any life critical process, having the appropriate personnel with the correct skills, mindset, and training can have a tremendous effect on the ultimate safety and outcome of a proposed job. There are several factors that interplay to ensure that the outcome of a potential job is as anticipated. Some of those factors include the appropriate choice of personnel, proper training and certification, team organization, supervision/leadership, and the correct use of rope access skills.

3-2 PERSONNEL QUALIFICATIONS

Not everyone is a good prospect to be a rope access technician. While there is no consensus regarding specific educational requirements that an individual must meet prior to being engaged in rope access training, they should, at a minimum, be

Professional Rope Access: A Guide To Working Safely at Height, First Edition. Loui McCurley.
© 2016 John Wiley & Sons, Inc. Published 2016 by John Wiley & Sons, Inc.

capable of reading and understanding product instructions, job hazard worksheets, rope access work plans, procedural manuals, and other materials related to safety and rescue equipment, methods, and training. They must also be capable of communicating effectively and cooperating well with coworkers.

Just as not every person is an ideal candidate to become a rope access technician, every rope access technician is not ideal for every job. Before one can determine who to choose for a specific job, a careful review of the proposed work should be conducted. Rope access is not, in and of itself, a pliable trade. Simply performing rope access does not directly produce an economic benefit. It is merely a method by which a laborer or an appropriately skilled tradesman or professional might reach a location to perform a task. Paramount, then, to the selection of personnel is the question of the task to be performed, whether it be painting, welding, cleaning, inspection, non destructive testing, or repair and maintenance.

In the case of skilled trades, it is generally far easier to add rope access skills to the toolbox of someone who already has appropriate skills, training and/or certifications required for the actual work than it is to add new trade requirements to an already certified rope access technician. Often, the skills required for a specific craft, such as welding or nondestructive testing, can take years to develop. While the skills and mindset required for rope access, particularly at the advanced supervisory levels also take time, generally, a skilled tradesman is a great asset even when certified at even the most basic levels of rope access (Figure 3-1).

FIGURE 3-1
Rope Access Certification Is Best Paired with Skilled Trades
© vuedici.org/BEAL.

Management and supervision of rope access is itself a specialty, and should be deployed only to experienced and competent persons who are able to demonstrate knowledge and abilities specific to management, leadership, and fall protection, in addition to possessing extensive rope access experience. In some cases, it is possible for an organization to outsource this role when internal resources are not available.

Aptitude for Rope Access

Working safely at height is more than just a matter of skills and knowledge; technicians must also exhibit suitable personality traits and an attitude that will contribute to appropriate and safe behavior while at height. Performing work while dangling in the air with a certain amount of air under one's feet is an acquired skill and can be disconcerting to the uninitiated.

Aptitude is not something that is easily measured, but a person's inherent suitability to rope access will either aid or hinder the ability to become proficient. A prospective rope access technician should be reasonably comfortable with, but not complacent regarding, the psychological effects of working at height. They should possess a nature that will not succumb to panic, should not be overly given to dramatic behaviors or foolish or undisciplined behavior, and should be capable of performing well in stressful circumstances.

Personal integrity is also an essential value for a rope access technician to possess. Technicians often work in remote, hard-to-reach locations and, as such, team members rely heavily on one another for safety and to ensure that the job is well done. A rope access technician must not be prone to shortcuts, but must be of a character to do the right thing, performing the task as they should, even (especially!) when they know that no one is watching.

While being a subjective consideration, the technician should possess a certain level of maturity to be able to maintain composure under unique working conditions, make decisions regarding life safety systems with competency, and take responsibility for themselves and others while working at height. This requires adequate skill and experience, as well as a certain ability toward multitasking. When a worker is performing a task at height, he must be cognizant of the rope access system as well as the work at hand. Constant attention must be given to both, in addition to environmental and other safety factors that may affect the work. Technicians must be capable of analyzing and recognizing hazards on an ongoing basis, acknowledging their own limitations, and reacting appropriately. Ego or a need to prove oneself has no place in rope access.

Aptitude and attitude may vary according to the environment (including height) of a particular work site or task. While a technician may be comfortable in a certain range of conditions and height, that same technician may not be appropriate for other environments.

Consideration of these factors can be a subjective undertaking, and requires a thorough review of credentials, previous experience, and exhibited behaviors. Consideration of previous rope access experience is useful, including follow up with references. Experience in recreational activities such as rock climbing, mountaineering, and caving may also be valuable, but weighed carefully to ensure that the candidate is suitably risk-averse. Experience with emergency services, technical rescue, or military forces may also be considered relevant, if these experiences involved working at or being exposed to heights.

The employer should make every effort to obtain references to verify claimed experience and competencies. Technicians can help maintain their reputation and assist a prospective employer in verifying their claims by maintaining a thorough recordbook that contains a log of hours and types of work at height, including names and contact information of employers and supervisors. Likewise, employers should be diligent in noting employee's work tendencies in their personal logbook so as to help promote appropriate use of logbooks. Where negative comments must be made, diligence should also be given to making amending notes to show progress on the part of the technician as appropriate.

Physical fitness of a rope access technician is also an important consideration. Again, specific requirements may vary depending on the particular job, including the specific skills being used, the frequency and duration of exercise over the course of the work, and potential hazards at the site. Personnel should be in general good health, and free from any disability that might prevent or interrupt working safely at a height. It is advisable for the employer to set baseline fitness criteria congruent with the type of work that is to be performed. At times, a rope access worker may need to temporarily refrain from working at height when taking certain medications or if other conditions exist that might affect the physical or mental capacity or behaviors.

3-3 TEAM ORGANIZATION AND COMPETENCIES

A rope access team should be fully self-sufficient, and properly supervised. Personnel should be selected for a given job based on their familiarity with the type of work to be performed, understanding of the workplace, understanding of the hazards and associated protective measures, and ability to work under the conditions likely to be present. Technicians must be trained to operate safely in accordance with established protocols as well as applicable regulatory requirements.

The Rope Access Plan should list members of the work team by name, along with their respective roles and responsibilities. A minimum team size consists of two technicians, one of whom is competent and designated as the supervisor; however, consideration must also be given to the minimum number of members required to perform the task, as well as to ensure that assistance is readily available to any work team member who needs it. For this reason, the composition and size of work teams will vary from job to job, based on the needs of that job as well as the skills and abilities of available technicians.

Where work is undertaken in a particularly hazardous or restricted environment, such as a confined space, or hazardous atmosphere, special consideration should be given to the training, abilities, experience, competence, and size of the work team to ensure that they are capable of dealing with any emergency that might arise from the work. If there is only one person on site who is fully qualified to perform rescue or retrieval, that person must refrain from placing himself in a position where he could become the one who needs rescuing.

Technician Skills and Responsibilities

Each member of the work team may have different responsibilities. There may be several rope access technicians at a given site who are assigned to carry out specific work tasks while engaging in limited methods of rope access. Workers should be focused primarily on their own safety, and secondarily on that of their co-workers. Levels of experience and requirements for hiring will vary depending on the particular application. Technicians assigned to be part of a work team may be certified at any level (most certification organizations use three levels), but they should all defer to the designated supervisor.

At the most basic level of certification, technicians should be trained not only to perform the required skills, but also to analyze the equipment and systems for safety during the course of work, and to work safely within the parameters of the Rope Access Work Plan.

A rope access technician should be adequately trained and have a thorough understanding of the following:

- Regulatory requirements applicable to the jurisdiction
- Means and methods for protecting oneself against falling objects
- Means and methods for protecting the public against falling objects
- Hazards involved with the use of work-tools
- Effective means of communication while on rope
- Analysis and protection against workplace hazards
- Purpose, establishment, and use of exclusion zones
- Implementation of a Rope Access Work Plan
- Purpose and use of a site-rescue plan
- Selection, inspection, and use and care of equipment
- Applicable recordkeeping requirements
- First aid for self and coworkers
- Simple rigging of anchors and vertical lifelines
- Use of a Rope Access Backup System
- Ascending and descending vertical ropes
- Ascending and descending through long and short rebelays
- Ascending and descending through a direction change anchor
- Other rope access skills as applicable to the worksite
- Coworker-assisted Rescue

Supervisor Skills and Responsibilities

Every distinct worksite must have a designated supervisor. There should be only one supervisor designated for every worksite. The assigned supervisor must be capable of implementing and overseeing an adequate system of work and safety for the assigned worksite. All work team members should work under the direction of, and defer to the supervisor, regardless of their own level of certification or experience – although any member of the work team should be encouraged to question matters pertaining to safety, and must have the latitude to not put themselves in a situation that they might feel is risky.

Specific knowledge and skill requirements for the supervisory role will vary from site to site. The individual assigned to supervise a given work should possess a good balance of sufficient knowledge and understanding of rope access along with a strong understanding of the work to be performed.

The Rope Access Supervisor should be fully capable of all the skills required for a technician, and should also possess additional training and skills related to systems rigging, advanced techniques, site management, and rescue.

> In addition to the previously outlined technician skills and responsibilities, a Rope Access Supervisor should be capable of the following:
>
> - Ensuring the safety and effectiveness of work methods
> - Ongoing monitoring and supervision of personnel and worksite
> - Preplanning for potential rescue needs
> - Ability to supervise and manage a rescue/incident
> - Ability to effect and perform complex rescues
> - Proper storage, care and maintenance of equipment
> - Development and use of a rope access plan
> - Ensuring adequate health and safety for employees
> - Hazard analysis and communications
> - Periodic detailed inspection of equipment, and associated recordkeeping
> - Monitoring and supervision of rigged equipment
> - Site access and safety control and management

Program Manager Skills and Responsibilities

Although the Rope Access Program Manager is often not part of the actual work team, it is not prohibited. Likewise, the Rope Access Program Manager may or may not also be the same person as the Fall Protection Program Administrator. The role of Program Manager is primarily a management function, helping to bridge the gap between administrative functions and field procedures. As such it may be helpful for the Program Manager to also be (or have been) a certified, competent rope access technician. The primary duties and responsibility requirements for this role include the following:

- Oversite of procedures for equipment selection, care and maintenance
- Oversite of the rope access Program
- Oversite of applicable risk assessments
- Maintenance of office records
- Maintaining records of accident and incident data

3-4 TRAINING AND CERTIFICATION

Training is an essential part of any work at height safety program, and in the case of rope access the training must include very specific knowledge and skills pertinent to methods for rope access. There is worldwide consensus on foundational requirements for rope access technicians, evidenced by multiple trade organizations who all follow a similar path, as well as an international ISO standard [1] on which numerous countries have collaborated.

All personnel involved in rope access operations should be trained in and capable of carrying out their assigned responsibilities with respect to the work being

---

[1] ISO 22846 Rope Access

TRAINING AND CERTIFICATION

performed specifically in relation to rope access. Beyond performing rope access techniques, hazard recognition and response, as well continuous attention to safety are also essential skills.

A training program should at a minimum include the following provisions:

- prerequisites (if applicable)
- training agenda and timeline
- learning objectives
- required training aids (manuals, equipment, audio/visual, and physical environment)
- student to instructor ratio specification
- methods of evaluation
- minimum performance requirements of students
- criteria for measuring success
- means for documenting successful completion

A thorough understanding of conventional approaches to fall protection, including fall arrest, positioning, and restraint, is essential for rope access technicians. Technician candidates should be familiar with common hazards related to work at height, particularly in relation to the industry and environment in which they work, and should be familiar with selection, care, maintenance, and retirement of PPE.

Rope access training should encompass methods that will ensure that the trainee gains the requisite knowledge and is also able to demonstrate the necessary skills to ensure safety in work at height. This includes the ability to demonstrate proper techniques, understanding of industry best practices, regulatory requirements and legislation in the jurisdiction in which they will be working, and familiarity and understanding of equipment.

Training Records

Technicians should maintain a thorough, documented record of all training and work experience, with their supervisor's verification details, and prospective employers should refer to and verify the accuracy and applicability of a worker's records before hiring them for a given job. Records may be kept in a written logbook or by electronic means, so long as the record contains sufficient detail for verification. An example of appropriate technician logbook information is shown in Figure 3-2.

Specific training is highly recommended prior to undertaking any actual work at height. Training may be provided by the employer, by a professional training entity, or some combination thereof; in either case, the knowledge and skills of the technician should be verified through evaluation by a competent person whose interests are commercially independent from the interests of the trainee. Periodic re-training and re-evaluation are appropriate.

Training Outline

All persons engaged in rope access work should be thoroughly trained on at least the following minimum requirements:

FIGURE 3-2
Typical Logbook. Courtesy of SPRAT.

Record of Rope Access Work Experience					
Date	Company or Organization	Details of Work Tasks	Location	Hours Worked	Supervisor Signature
Running total of hours worked					

1. Review of common methods and means of conventional fall protection.

This topic might take as little as an hour for students who already have a good foundation, or up to several hours for students who may not already possess knowledge or experience in this area. Understanding conventional methods of protection is essential for the rope access technician because of the frequency in which a given jobsite will employ multiple methods simultaneously or consecutively. If jurisdictional requirements already exist for basic training of persons who work at height using conventional fall protection, requiring this training to have been completed as a prerequisite to rope access training is preferable. However, even in the case of students who have pre-existing knowledge and/or experience, a thorough review is recommended to ensure a common baseline. Chapters 1 and 2 of this book may be used to achieve this goal.

2. Introduction to rope access concepts, principles, and use

Many people still believe that rope access is simply rappelling with fall protection as backup. For this reason, the training program should thoroughly review the basic concepts, including that of continuous double protection, interchangeability of systems, and provision for rescue. The difference between rope access backup systems as compared with fall arrest should be clearly delineated, and clarification provided regarding the requirement for a rope access program to embrace a complete system of equipment, methods, and verified skills. This is a good time to review the Rope Access Permit/Plan concept, and examples may be shared of how rope access is used in different industries. At least two hours should be spent on this topic, and perhaps more in the case of more advanced students who may be on a path toward becoming supervisors or program administrators. Chapters 2 and 3 of this book may be used to support this part of training.

3. Rigging Equipment selection, inspection, use, maintenance, and retirement

Students should be familiar with equipment used for rigging systems, and should have a good understanding of both performance and regulatory requirements. Foundational system components such as anchorage connectors, ropes, and connectors are often shared among many users in a workteam or on a job, while items such as ascenders, descenders, lanyards, backup devices and similar equipment that is used directly by the technician may be more appropriately assigned to individual persons. It should be acknowledged that the lines between what components of equipment are shared rigging gear versus that which is more often considered to be Personal Equipment can be blurry. Drawing clear lines of distinction is less important here than establishing a good understanding of criteria for selecting equipment for different types of uses. In the case of rigging equipment, a more thorough understanding of component rigging analysis, failure modes, and behavior under load is required. Chapter 4 will provide an excellent overview, and may be covered in as little as an hour or two, but the concepts surrounding appropriate inspection and use should be considered throughout the course of the training.

    4. Personal Equipment selection, inspection, use, maintenance, and retirement

A follow-on to the rigging equipment discussion, Personal Equipment includes that gear which technicians often prefer to keep and maintain on their person. This topic is discussed in detail in Chapter 5 of this book. Any work at height requires careful attention to selecting the proper Personal Equipment for the job, knowledge of how to perform preuse inspection, understanding of what constitutes appropriate use, and information on expiring equipment. A subcategory of Personal Equipment, PPE is equipment worn to minimize exposure to a variety of hazards.[2] PPE for different types of work at height will have varying requirements, and therefore in the training phase it is important to teach the worker how to select the equipment rather than to specify what equipment should be used. While an introduction of an hour or two may suffice to provide an overview, the concepts of inspection and proper use should be woven throughout the entirety of the training event. Students should be expected to inspect their equipment daily and prior to use, and should thoroughly understand the capabilities and limitations of the particular components that they are using. Chapter 5 of this book provides a good baseline for instruction, but emphasis should be placed on the hands-on aspects of this training.

    5. Basic systems rigging concepts (forces, angles, and rope management)

Even at the most basic level, a rope access technician should be capable of building basic access and backup systems as well as analyzing any system they are about to use – even if it was initially built by someone else. A basic understanding of how forces behave in a system is essential to this ability. Understanding a system will help the technician to use it most efficiently and safely. This part of training should cover the effects of edges, dynamic loading, swing fall, friction, angles, deviations, rope management, forces, and other related factors that can affect system performance. Again, up to a half day can be spent to focus on these concepts, although this time need not be consecutive. In addition, these principles can and should be interwoven through the entirety of the training, as they are perhaps the most important bit of foundational knowledge a technician should have. Chapter 6 will support this part of the training.

    6. Anchor selection, rigging, and analysis

Anchorages form the foundation for every work at height system, and there are several basic principles that apply to the selection of anchor points and rigging of anchor systems. A good understanding of how forces, directional loading, and impact loads can potentially affect the anchorage is important for every person who may work at height. As much as a half day can easily be spent on anchorages, although once again this is primarily a hands-on skill that should be addressed with every system that is used during the course of the training event. Chapter 7 will be of use in training these concepts.

    7. Rigging and use of a Rope Access Backup system

While the rope access and backup systems along with the associated equipment, are generally thoroughly outlined in the Rope Access Plan developed by the

---

[2]U.S. Department of Labor; Occupational Safety and Health Administration; Information Booklet 3151-12R, published 2003

Program Manager and implemented by the Supervisor, technicians at every level should have a good grasp of knowledge regarding the rigging and use of the backup system, simply because their lives depend on it. While the work is directed and overseen by the supervisor, technicians may be involved in initial rigging of the system and certainly use it on an ongoing basis. Preuse inspection is the responsibility of every technician, as is the ability to use the system properly and within its limitations, and to recognize when something is wrong. Chapter 8 provides an overview of the rope access backup system, and Chapters 9–13 explore the use of the backup system relative to specific access techniques.

8. Rigging and use of a Descending and Ascending System

The same principles mentioned regarding the backup system also apply to the primary system. Although not every technician is always involved in rigging the system, they are most certainly using the system throughout the course of the work, so a thorough understanding of its capabilities and limitations is essential. Every technician must also be capable of preuse inspection, using the system properly and within its limitations, and recognizing when something is amiss. Descending Systems are discussed in detail in Chapter 9 while Ascending Systems are discussed in Chapter 10.

9. Changeovers

Changing over from ascent to descent, and vice-versa, form the basis for all other skills that involve passing obstructions, moving between systems, and performing rescues. The duality and interchangeability between the primary and backup systems is foundational to rope access, and the ability to comfortably interchange between these systems is precisely what gives the rope access technician such versatility and a depth of safety. This skill is foundational to the versatility and 100% protection that rope access demands, and may be found in Chapters 9 and 10.

10. Passing an intermediate anchor

Passing an intermediate anchor, also called a rebelay or re-anchor, is a common practice on long drops or climbs, or anywhere where multiple anchor points are used to prevent swing fall, danger from dropped objects, or other hazards. It is considered to be a foundational skill, and guidance for performing this maneuver is found in Chapter 11.

11. Passing a deviation

Deviations are intermediate points where a rope is redirected for the purpose of changing the fall line or direction of travel for the technician. Passing a deviation is not rocket science, but requires certain provisions and considerations for efficiency and safety, and the ability to go back in whatever direction the technician has come from. This, too, may be found in Chapter 11.

12. Use of Powered Devices in Rope Access

Not every rope access project or site involves the use of powered devices, but because these are becoming increasingly common it is highly recommended that technicians have at least a baseline familiarity with the capabilities as well as the limitations of their use. An overview of powered devices is offered in Chapter 12.

13. Vertical rescue by lowering

A technician must be capable of performing rescue. Rescuing another without entering a fall hazard oneself is always the preferred method whenever feasible. Rescue by lowering can allow a basic technician to rescue a more advanced technician, even if the lesser trained technician is not capable of or comfortable with performing the rope access technique that the rescuee is in. Guidance for rescue methods is provided in Chapter 13.

14. Vertical rescue by descent

Most rope access work involves simple vertical systems. Having the ability to rescue oneself, as well as another, from such a basic system is the most essential form of rescue. This type of rescue puts the rescuer into the system and is performed when a rescue by lowering is not feasible. It is discussed in Chapter 13.

15. Basic first aid

It is highly advisable for all workers to possess at least a rudimentary knowledge of CPR and first aid so that they can assist their coworkers when necessary. No one wants to face an agonizing situation of being at the site of an injury or illness and not knowing what to do. Some parts of regulation, such as the OSHA confined space regulation, require that at least one responder be certified in first aid and CPR. This book is not a medical reference, so additional training should be sought for basic first aid/CPR. Suggested resources include the **American Red Cross**, the **American Heart Association**, the **National Safety Council**, the **National Ski Patrol**, etc. The employer's written emergency action plan should provide more thorough guidance for recommended actions and responsibilities in the event that a worker is injured.

Training is not complete without an effective evaluation process to ensure that the trainees have captured key information from their training, and to verify that they are capable of performing the requisite skills. Although an evaluation can only measure the ability of the trainee at the time and place of evaluation, and does not assure long-term retention of the information, it at least confirms a baseline of knowledge. During evaluation, trainees should exhibit an understanding of information through verbal or written responses to pertinent questions, and they should be capable of competently demonstrating all of the necessary skills for their respective level of certification in a safe and timely manner.

Upon proving their capability in all of the above requirements, technicians may be adequately prepared to work under the direction and supervision of a more experienced rope access technician. However, completion of all of the above does not by itself constitute qualification to supervise or to work without direct supervision.

An experienced trainer may be able to cover this material including the independent evaluation in as little as 40 hours. However, the duration of a training program should not be considered to be the key parameter. Ensuring that the student comes away from the training with the requisite knowledge and skills is most important. To ensure adequate knowledge, evaluation of a candidate's knowledge and skills should be performed by an evaluator who is independent from the training entity.

Advanced Levels of Certification

Most Rope Access Certification Bodies recognize three distinct levels of certification for rope access technicians. At all levels, technicians should be capable of all of the aforementioned up to some level. The distinction between levels is generally based on experience and greater skills in rigging and rescue. Figure 3-3 summarizes the essence of these differences, while Chapter 21 offers greater detailed guidance on the level of competence normally expected for technicians certified to each of these three levels in accordance with the internationally recognized ISO 22846-2:2012 standard.

FIGURE 3-3
Technician Competency Levels as Shown in ISO 22846.

Basic	Intermediate	Advanced
• Ability to evaluate a system • Elementary Rigging • Equipment inspection & Use • Simple Maneuvers • Use of rope access backup • Use of work plan • Simple Rescue	• Competency in all Basic requirements as a prerequisite • Verified field experience • Advance Rigging • Difficult Maneuvers • Difficult Rescue • Basic Supervisory Skills	• Competency in all Basic & Intermediate requirements as a prerequeisite • Verified extensive field experience • Team direction • Complex Rigging • Complex Rescues • Site Management

Rope Access Certification Bodies

Prior to hiring a rope access technician, proof that they possess current certification by an appropriate Rope Access Certification Body should be ensured. This may be provided in the form of a certificate or a card, as shown in Figure 3-4. A copy of the certificate or card should be retained by the employer, and certification may be verified through the certifying organization.

Certification as a rope access technician is only as good as the organization bestowing that certification. While the vast majority of universally recognized Rope Access Certification Bodies are nonprofit membership based organizations who represent the technicians themselves and/or rope access employers, a few privatized entities may be found who claim to be capable of bestowing certification.

Suffice to say that the technician and employer should approach certification by privatized entities with some caution. Private corporations are, by definition, motivated by a desire to gain profit and otherwise inure to the benefit of the owner(s). Member-based organizations, on the other hand, are typically managed

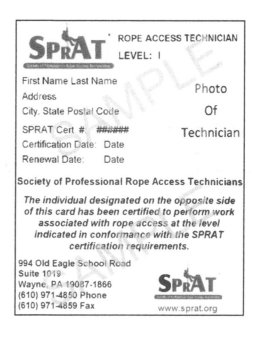

FIGURE 3-4
A Typical Certification Card. Courtesy of SPRAT.

and overseen by a board elected by their members, arguably creating a more balanced representation of industry best practices.

Organizations that offer certification services should do so in accordance with ISO 22846. Some of the organizations throughout the world that currently offer certification that is in compliance with ISO 22846 include the Society of Professional Rope Access Technicians (SPRAT), Australian Rope Access Association (ARAA), Singapore Rope Access Association (SRAA), Industrial Rope Access Trade Association (IRATA), etc.

3-5 SUMMARY

It is the responsibility of the employer to ensure that a selected employee is adequately trained and is capable of performing the required level of work. It may be necessary for technicians who are not continuously using their rope access skills to undergo refresher training. Because rope access requires a combination of physical skills and mental aptitude, such refresher training may need to include conditioning for exposure to height as well as specific access skills and methods.

As a technician gains experience and knowledge, and progresses to greater understanding of rope access systems and its use, it may be appropriate for them to seek additional knowledge and skills in more advanced methods used for access in complex environments, more advanced supervisory skills, and the rescue capabilities to match. This book also provides sufficient information to support advanced training such as this.

CHAPTER 4

Equipment for Rigging

Tom Wood

By the end of this chapter, you should expect to understand:

- How to choose equipment for rigging in life safety situations
- Equipment considerations for rope access
- How PPE fits into the broader category of Personal Equipment
- The difference between Personal Equipment and Team Equipment
- The importance of using only purpose-designed equipment for life safety
- How to approach regulatory considerations
- How to interpret strength ratings of equipment
- Factors that affect selection of equipment
- Additional considerations when using equipment for rescue
- How to choose between different types of life safety rope
- The important considerations for anchorage connectors
- The basic requirements for carabiners and other connectors
- Some important facts about hardware

4-1 EQUIPMENT FOR RIGGING ROPE ACCESS SYSTEMS

Clean, safe, and efficient rigging in rope access is a blend of art and physics. While many of the rigging principles and methods used by early Egyptians are still used today, technological advances in rigging equipment have gone a long way toward increasing both efficiency and safety margins. Indeed, the whole notion of safely and efficiently executed suspended rope access work is heavily dependent on the rigging knowledge and expertise of the rope access technician. So, whether a rope access technician is setting up a simple rappel to inspect a building façade or rigging a complex, or a multipoint load sharing anchor to support the weight of a sign weighing hundreds of pounds, having a thorough understanding of rigging tools and techniques is essential.

Professional Rope Access: A Guide To Working Safely at Height, First Edition. Loui McCurley.
© 2016 John Wiley & Sons, Inc. Published 2016 by John Wiley & Sons, Inc.

Rope access is not a trade unto itself, like that of electricians, welders, or iron workers: Rope access is more like the bus you drive to get to work.[1] So, when discussing the equipment used for rope access rigging, it's important to recognize the makes and models of the other "buses on the road" to fully understand how the equipment and techniques used for rope access can either complement or contraindicate use in other methods of access like fall arrest, cranes and derricks, suspended scaffolds, and rescue.

4-2 HOW TO CHOOSE EQUIPMENT FOR RIGGING IN LIFE SAFETY SITUATIONS

A well-designed and meticulously constructed rope access system's roots can be traced directly back to the individual pieces of equipment selected for the task at hand. Although the immutable laws of physics have the final say when it comes to how well a system functions, the local regulatory concerns that govern the plot of land on which a system is built may also dictate what equipment is used (or not used) in any given rope access system.

In some countries or municipalities, the equipment used for rope access may not meet local fall arrest (or related work at height methods) equipment standards. In these instances, the technician relies on what is known as an "industry best practice" to justify the selection and use of certain PPE and rigging equipment. Some employers address these regulatory concerns by writing a company risk assessment that justifies the use of a particular piece of equipment.

It is important for each technician to understand the local rules and regulations that govern a jobsite, and to have an awareness of the instances when there is a conflict between regulatory requirements and rope access industry best practices regarding equipment selection. For example, in the United States, the Society of Professional Rope Access Technicians (SPRAT) states in their Safe Practices for Work document that locking connectors with a major axis strength of 22 kN (5000 lbs) are acceptable for use when performing rope access.[2] However, according to the American National Standards Institute (ANSI) Z359.12-2009 Fall Protection Code,[3] when connectors are used for fall arrest in the United States, they should be *autolocking*, with a major axis strength of at least 22 kN AND have a *gate strength* of at least 16 kN (3600 lbs). In addition, the ANSI Fall Protection Code specifically disallows two carabiners to be connected to each other,[4] which is a common practice for most rope access technicians. Although many might argue that true rope access should be categorized somewhere in between work positioning and work restraint (not fall arrest), the discrepancy between the two organizations' equipment regulations remains.

However, ultimately, irrespective of work at height rules and regulations that govern a jobsite where the work is being performed and the regulatory requirements

---

[1] Loui McCurley, overheard in conversation.
[2] The Society of Professional Rope Access Technicians, *Safe Practices for Rope Access Work*, August 2, 2012 edition, (2.9) 4, (10.4) 14.
[3] American National Standards Institute (ANSI) *ANSI/ASSE Fall Protection Code, Version 3*, 2012, Z359.12, (3.1.1.3) 9–10.
[4] American National Standards Institute (ANSI), *ANSI/ASSE Fall Protection Code, Version 3*, 2012, (7.2) 56.

FIGURE 4-1
Most life safety equipment used in rope access will be marked with a Minimum Breaking Strength {MBS} or Safe Working Load {SWL}. Courtesy of Pigeon Mountain Industries, Inc.

and best practices regarding equipment selection, it goes without saying that ALL rope access equipment should be selected, used, inspected, maintained, marked, stored, and retired in a manner consistent with the recommendations of the manufacturer. So when choosing rigging equipment for life safety applications, a combination of function and application may be called for.

> When choosing rigging equipment for life safety applications, a combination of function and application may be called for.

Component strength, along with regulatory considerations, is one of the primary concerns when choosing rigging equipment. Most manufacturers provide either a Working Load Limit (WLL), a Safe Working Load (SWL) or a Minimum Breaking Strength (MBS), as shown in Figure 4-1, with any piece of life safety equipment used for work at height.

This manufacturer-provided information is critical when selecting equipment for rigging, since both the WLL and/or the MBS for each individual piece of rigging equipment can be used to evaluate the limitations of use for that component and, in turn, the system. Of the two, the more essential piece of information for the technician is the MBS. Knowing the strength of a component allows a competent rigger to analyze system safety factors, whereas the WLL alone provides little insight to the technician.

Along with strength, efficiency is also important when selecting equipment for rigging. A piece of gear that is strong enough to support the weight of a battleship is of little use if its inclusion detracts from efficiency or ease of use within a system. Though pulleys often come to mind when thinking of equipment efficiency within a system, there's more to it than just the amount of efficiency lost to friction (as is the case with pulleys)-applicability and ease of use for the job at hand is also an important consideration. Gear selected for rigging use in a rope access system needs to match up with its intended use. For example, if the mechanical advantage system being used involves the use of tandem prusiks for progress capture, it might be wise to purchase prusik-minding pulleys to accommodate this method.

4-3 THE DIFFERENCE BETWEEN PERSONAL EQUIPMENT AND RIGGING EQUIPMENT

Generally speaking, the term *Personal Equipment* refers to equipment worn or used directly and individually by the technician. More specifically, *Personal Protective Equipment* (*PPE*) is a narrower category within the context of Personal Equipment that can be thought of as anything worn by the rope access technicians to protect their health or safety. In other words, when the elimination of a hazard isn't possible, PPE can help mitigate the effects of that hazard on the technician. *Rigging equipment*, on the other hand, refers to equipment that is used to create the working systems (the primary system and the backup system, as discussed in Chapter 8) that, when used appropriately along with Personal Equipment, can mitigate potential work at height hazards.

> Personal Protective Equipment (PPE) is usually something worn by the technician; Rigging Equipment is typically something used by the technician.

The broad category of rigging equipment may include both Team Equipment and Personal Equipment. Equipment must be used properly and in harmony to maximize safety. So for instance, until zero-gravity harnesses are available, the immutable laws of gravity will require that PPE be worn in conjunction with an appropriately rigged system to mitigate the potentially fatal effects of gravity (in other words, a fall).

In summary, Personal Equipment is usually something *used individually* by the technician, and rigging equipment is typically something that technicians use for *rigging systems*.

This chapter will focus primarily on those components of rigging equipment that are commonly shared among users on a worksite – that is, equipment used to create the systems that rope access technicians use. Chapter 5 will address Personal Equipment, including PPE as well as Personal Equipment that a technician will generally own and carry as part of their personal kit.

> Personal Equipment is usually something used (and often owned) personally by the technician, and rigging equipment is typically something used and shared on a rope access site.

Anchorage connectors, connectors, slings, rope grabs, pulleys, rope, rope protectors, and rope grabs would all be considered rigging equipment.

4-4 RIGGING EQUIPMENT FOR FALL ARREST

Fall arrest and rope access have always had a love/hate relationship. However, like a bickering couple that can't ever seem to part ways, both rope access and fall arrest need each other. Much of the confusion surrounding the use of the rigging equipment used for both rope access and fall arrest comes from the common

FIGURE 4-2
Fall arrest and rope access equipment, regulations and techniques are often quite different. Courtesy of Vertical Rescue Solutions by PMI.

misconception that the two methods of access are interchangeable. While they do share some similarities (rope, connectors, and anchorages are required for both types of systems), regulatory agencies look at each in very different terms as far as the gear that is used.

The ANSI Z359 Fall Protection Code states the differentiation well: "Rope access is different from fall arrest, fall restraint and other fall protection techniques in that the authorized person is generally fully suspended by the rope system during work. The safe use of rope access systems requires specific competence in rope access techniques acquired by training and experience, confirmed with independent assessment and certification by one competent to assess and certify rope access skills and knowledge."[5]

Since technicians often use fall arrest gear and techniques when working near an edge or accessing their anchors near an edge, they need to be sure that they are using equipment that meets the demands and regulatory requirements of fall arrest when using fall arrest methods for protection. Equipment used for fall arrest, as shown in Figure 4-2, often involves very different techniques and regulations than does equipment used for rope access.

4-5 RIGGING EQUIPMENT FOR CRANES VERSUS RIGGING USED IN ROPE ACCESS

When using *cranes or derricks* for lifting, there are often strict federal or municipal regulations that dictate what equipment and techniques can be used. In the United States, crane operators, riggers, and signal persons look to OSHA, CFR 1926, Subpart CC, Cranes and Derricks in Construction, for guidance on rigging, lifting, and certification levels when using cranes.

Rigging equipment for a lift with cranes, such as that shown in Figure 4-3, is very different than the equipment used by technicians to lift a load using rope access techniques.

Primarily, cranes and derricks use wire rope and cable in place of synthetic ropes or slings. In addition, cranes have very strict regulations and guidelines (which vary

---

[5] American National Standards Institute (ANSI) *ANSI/ASSE Fall Protection Code, Version 3*, 2012, Z359.0 2012, E2.149, 26–27.

FIGURE 4-3
Cranes have completely different rules and regulations than rope access governing the selection and use of rigging equipment for lifting. Courtesy of Vertical Rescue Solutions by Pigeon Mountain Industries, Inc.

from industry to industry) when it comes to lifting human loads on the jobsite. So when comparing rigging equipment for cranes to the rigging equipment for rope access, there are very few similarities, and they are NOT interchangeable.

4-6 RIGGING EQUIPMENT FOR SUSPENDED SCAFFOLDS

Suspended scaffolds, introduced in Chapter 1, may be mixed and matched with rope access methods to get a job done. Most suspended scaffolds rely on a mechanized method of raising or lowering the platform via ropes or cables. Ropes and cables used for scaffold suspension should not be used interchangeably for rope access, and vice-versa.

Often, workers using suspended scaffolds or swing stages rely on a rope grab connected to their dorsal point and attached to a laid or braided rope for their safety line, as shown in Figure 4-4. These rope grabs are not compatible with the fall arrest rope grabs used on kernmantle ropes.

4-7 RIGGING EQUIPMENT FOR RESCUE

When it comes to the selection and use of equipment for the *rescue* of an injured or stranded coworker, the technician has a lot of important decisions to make. Does the

FIGURE 4-4
Conventional Fall Protection systems often use laid or braided rope. Courtesy of Reliance Industries, LLC.

subject needing rescue need to be raised, or lowered? Will there be a need to put a "two-person" load on the device? Is the gear being used for the rescue rated for a "two-person" load? What environmental concerns affect how rescue equipment can be used?

Equipment used for rescue after a fall may differ from that used for rope access, as shown in Figure 4-5, particularly if the rescue is not being performed using rope access rescue methods. It's important to note that equipment for at-height coworker rescue usually falls under a different regulatory category than for an at-height rescue carried out by professional rescuers (such as the National Fire Protection Authority (NFPA). In addition, rope access equipment that is suitable for on-rope, day to day work may not be appropriate for the demands of a rope access rescue. If rescue equipment for a rope access rescue is different than that used for rope access work, this should be noted in the Rescue Preplan and in the Jobsite Hazard Analysis (JHA).

4-8 ROPE

Not all ropes are created equal, and the savvy technician knows the importance of matching the right rope with the right job.

Primarily, the ropes used for rope access are of the kernmantle construction type, usually between 10 and 12.5 mm in diameter and certified for life safety use.

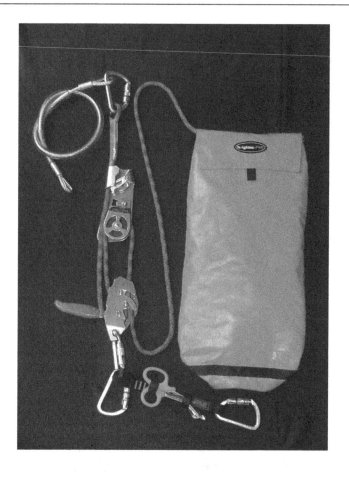

FIGURE 4-5
Equipment used for rescue may differ from the equipment used for every day work. Courtesy of Pigeon Mountain Industries, Inc.

Kernmantle ropes are manufactured with an inner core (kern), which is responsible for most of the rope's strength, and an outer sheath (mantle) which protects the core.

Though there are a variety of man-made materials used for rope construction, most ropes are made of nylon and/or polyester. Different fibers offer different performance characteristics, as shown in Figure 4-6. Natural fibers, such as manila and sisal, are less strong to begin with, and deteriorate over time. High modulus fibers, such as HMPE's and *para*-aramids, are of very high strength but do not offer much forgiveness in the area of shock absorption. Nylon and polyester offer a good balance of strength, force absorption, and chemical resistance. Between the two, nylon tends to offer slightly better shock absorption than polyester, but absorbs water at a molecular level, which can lower a rope's MBS by roughly 10% (the nylon rope regains its full strength when dry). Polyester ropes are more hydrophobic than nylon ropes, but can sometimes feel stiffer after being loaded.

Compatibility between ropes and the equipment used on them (e.g., rope grabs, descenders, ascenders, etc.) is often a key consideration when determining what rope to choose for a particular application.

The term compatibility encompasses several concepts, some of which might be rather subjective. The simple parts of compatibility involve such concepts as dimensional suitability. For example, most descenders and rope grabs are designed to function with a limited range of diameter of ropes. This is usually noted on the device. The subjective criteria require more experience to analyze. For example, some descenders have a tendency to milk the sheath of a rope, and some ropes are more susceptible to milking than others. These two characteristics together would be considered "incompatible."

Fiber performance characteristics				
	Breaking tenacity (gpd)	Chemical susceptibility	Melting temp (°C)	Elongation at break (%)
Manila	5–6	Acids, alkalis	Chars @ 148	10–12
Sisal	4–5	Acids, seawater	Chars @ 148	10–12
Nylon	7.5–10.5	Mineral acids	218–258	15–28
Polyester	7–10	Sulfuric acids, alkalis	254–260	12–18
HMPE	25–44	Minimal	144–155	2.8–3.9
Para-Aramid	18–29	Strong acids and bases	Decomposes @ 500	1.5–4.4

FIGURE 4-6
Performance comparison of Fiber Rope Characteristics.

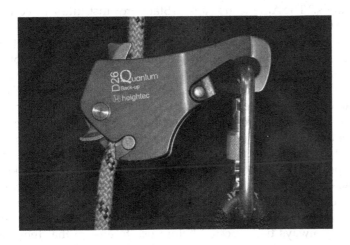

FIGURE 4-7
Rope and equipment compatibility is key. Courtesy of Pigeon Mountain Industries, Inc.

Compatibility refers to the idea of two or more components being well suited to be used together, as shown in Figure 4-7. It is impossible to describe all of the subjective considerations that should be taken into account in determining compatibility. These are appropriately left to the discretion of the trained and experienced competent person. It is only through a combination of knowledge and experience that a person can become competent to make such determinations.

There are three commonly used types of kernmantle life safety ropes: *Static*, *Low Stretch*, and *Dynamic*.

- Static rope, most commonly found in the United States, is most often used when little to no elongation is desired (such as use as the main line, or for highline use). Static rope, as defined by the Cordage Institute, has less than 6% elongation at 10% of its MBS. For example, a 11 mm static rope normally has a breaking strength of at least 6000 lbs (26.7 kN). To be considered a static rope, this rope would need to have less than 6 ft elongation per 100 ft of rope when loaded to 600 lbs.
- Low stretch rope, per Cordage Institute standards, has between 6% and 10% elongation at 10% of its MBS. Rope meeting European (EN) standards is usually low stretch (aka semi-static) rope. Compared to static rope, a low stretch rope has lower impact forces on both the anchor and the technician in the event of a shock load. This characteristic makes it a good choice for a safety line.

- Dynamic rope is most commonly used for recreational climbing, but it does have a place in industrial rope access if used in the right situation. Wherever there is a greater than Factor 1 Fall (or when the technician is climbing above an anchor) the use of dynamic rope might be called for to limit the impact forces generated in a fall. Elongation of dynamic rope is calculated differently than it is for static and low stretch rope. Tensile (static) loading is applied and measured with static and low stretch ropes. Dynamic rope is quantified by its ability to absorb force in a significant shock load and is rated for its ability to withstand a number of repeated shock loads of a specified magnitude. Many rope access lanyards (cow's tails) are made of dynamic rope.

Life safety ropes have several industry standards to which they can be certified, both in the United States and in Europe. The Cordage Institute standard CI-1801 addresses life safety rope and requires 7/16-in. (11 mm) rope to have a MBS of at least 6000 pounds.[6] The American National Standards Institute sets requirements for synthetic rope lifelines used in personal fall arrest systems.[7] The European Standard EN 1891 classifies ropes as either Type A or Type B. Type A ropes are suitable for rope access work, whereas Type B ropes are required to have a strength of only 18 kN (4047 lbs) and should not be used for rope access work.[8]

The use of properly applied swages, sewn loops, or properly tied knots are all considered acceptable practice for creating terminations in rope for rope access. Terminations are addressed in greater detail in Chapter 8. Factory terminations tend to retain more of the rope's strength, but may lack the shock absorbing capabilities of a knot used for termination, as shown in Figure 4-8. Generally speaking, when a knot is tied in a rope, it can weaken the rope by 20–30% (but based on the knot selected and the radius of the bend where a knot is tied, the strength reduction could be higher). A rope termination shouldn't reduce the overall strength of the rope below the necessary system strength required to achieve the prescribed safety margin.

Ropes must always be inspected before use to ensure that they are in good and usable condition. Proper rope inspection requires objective and subjective

FIGURE 4-8
Knots generally lower the breaking strength of a rope by 20–30%. Vertical Rescue Solutions by Pigeon Mountain Industries, Inc.

---

[6] Ibid.
[7] *Safety Requirements for Personal Fall Arrest Systems, Subsystems, and Components.* ANSI/ASSE Z359.1:2007.
[8] *Personal protective equipment for the prevention of falls from a height. Low stretch kernmantle ropes.* European Standard EN 1891:1998.

CONNECTORS

evaluation by an experienced technician. It will include both a visual and a tactile inspection, and every inch of rope must be inspected.

4-9 CONNECTORS

Any removable piece of hardware that connects parts of a system or subsystem together may be classified as a connector. Whether a connector is a pear, oval or offset D- shaped *carabiner*, a *screwlink*, or a *snaphook*, connectors are an integral component of any rope access system. Connectors come in all shapes and sizes, as illustrated in Figure 4-9, and their intended use dictates which type is suitable for the job.

Although connectors are usually constructed of aluminum alloy or steel, determination of which connector is best suited for a given job depends largely upon the environment and intended use of the connector.

The most commonly used type of connector used for rope access is the carabiner, as shown in Figure 4-10.

Carabiners used in rope access should be of a self-closing, locking design and of sufficient strength to meet applicable regulations in the jurisdiction where the work is being performed. Recreational carabiners are generally not appropriate for rope access work.

Performance characteristics of a carabiner are derived in part from the material from which it is made, and in part from its geometric dimensions and design. Rope access technicians are primarily concerned with the strength of the connectors, and with their compatibility with other components.

Carabiners may be made of aluminum alloy or steel and should be rated to a MBS of at least 22 kN (5000 lbs) along the major axis. In the United States, there has been a recent move toward ANSI Z359 compliance for connectors. This standard requires a MBS of 5000# in the long axis and not less than 3600# in the short axis direction. Figure 4-11 illustrates the parts of a carabiner, including the long and short axis.

Generally, D- and Offset-D-shaped carabiners offer the best strength-to-weight ratio, and as a result many carabiners selected for rope access are of this shape. The D shape also helps to align the load along the strongest side of the carabiner, the spine. Carabiners are strongest when loaded in this configuration.

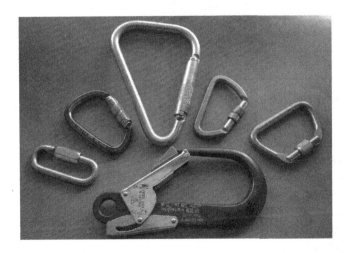

FIGURE 4-9
Connectors come in all shapes and sizes. Vertical Rescue Solutions by Pigeon Mountain Industries, Inc.

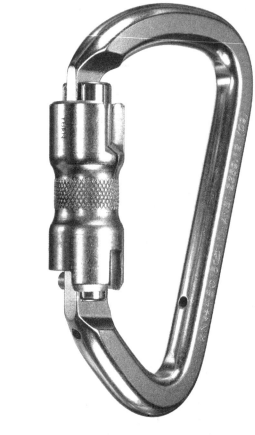

FIGURE 4-10
Self-closing, locking carabiner used for rope access. Courtesy of SMC – Seattle Manufacturing Corporation.

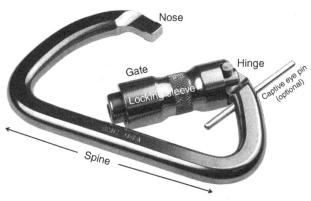

FIGURE 4-11
Parts of a carabiner.

Selection criteria for carabiners

- strength (major and minor axes)
- double-action opening mechanism
- locking mechanism, as appropriate
- ergonomic function
- dimensions
- material
- suitability for use with gloved hands
- compatibility

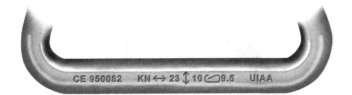

FIGURE 4-12
Connector markings. Courtesy of Pigeon Mountain Industries, Inc.

Carabiners should be marked to indicate the name of the manufacturer, certification organization, strength, and any standards that the carabiner meets, as shown in Figure 4-12.

Most carabiners that are used in rope access feature a locking gate, which contributes to both strength and security. Carabiner locking mechanisms may be of a screw-lock or a self-locking type. Manually locked carabiners require the user to engage a locking mechanism to keep them from opening accidentally. Autolocking connectors require at least two actions to open them, and they close and lock automatically when the gate is released by the user. For compliance with the aforementioned ANSI Z359, carabiner gates must be of a self-locking design.

Most rope access organizations do not allow the use of nonlocking connectors, as shown in Figure 4-13, in a life safety application. They should not be a part of any rigging intended to support a human load, or in a situation where injury or loss of life would be the result of a system failure. In addition, linking two or more nonlocking connectors together can cause the connectors to come undone if a twisting force is applied during loading. This is known as dynamic rollout, and it's been known to cause accidents and fatalities.

Connectors must be compatible with the connection to which they will be attached. Compatibility refers to the relationship between multiple components. It is not possible to select one carabiner that will always be compatible with every possible connection, so the competent user must exercise judgment about this in the field. Shape, size, and gate design should allow the carabiner to maintain position in its connection without jamming, loosening the connection, or being susceptible to rollout as described above.

Screwlinks (aka maillons) are manually locking connectors that are typically smaller than carabiners. They are most appropriately used in Personal Equipment applications, as described in Chapter 5.

Only screwlinks that are specifically designed for life safety use, such as that shown in Figure 4-14, should be used for rope access. Care should be taken to avoid the use of hardware store (commodity) screwlinks in life safety applications, since they are not man-rated and have WLLs below those acceptable for life safety use.

Snaphooks, shown in Figure 4-15, are a type of connector commonly found in fall arrest applications. These are unique and differentiated from carabiners by geometry, dimension, and use. Snaphooks generally comprise a hook-shaped member with a normally closed keeper, which may be opened to permit the hook to receive an object and, when released, automatically closes to retain the object. Snaphooks are often found at the ends of fall arrest or work positioning lanyards, and are particularly susceptible to misuse, including rollout. They should always be autoclosing and autolocking, and always be attached to a connection point suitable for their intended use. The user should avoid connecting snaphooks to each other, to a horizontal lifeline, or to any object, which is incompatibly shaped in relation to the snaphook such that the connected object could depress the snaphook keeper and release itself.

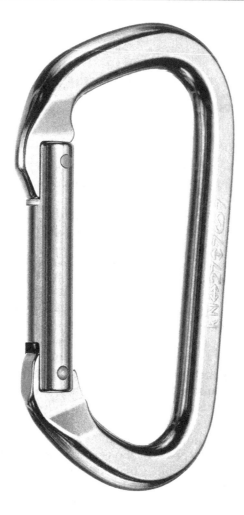

FIGURE 4-13
Nonlocking connectors should not be used for life safety applications in rope access. Courtesy of SMC – Seattle Manufacturing Corporation.

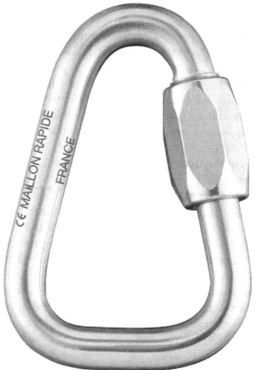

FIGURE 4-14
Screwlink. Courtesy of EDELWEISS SAS.

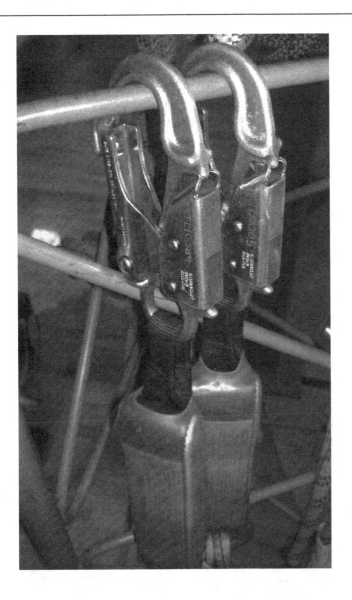

FIGURE 4-15
Snaphooks are often found at either end of fall arrest lanyards. Vertical Rescue Solutions by Pigeon Mountain Industries, Inc.

4-10 HARDWARE

When it comes to rope access rigging equipment, the term *hardware* can cover a broad range of gear. Rigging plates, pulleys, ascenders, descenders, and rope grabs would all fall under this category. While some hardware for use in rope access is relatively low-tech and simple – consisting of a single piece of stamped or molded metal – other types of hardware are complex, high-tech devices made of many different materials and containing lots of moving parts.

- Rigging plates, shown in Figure 4-16, are usually solid pieces of aluminum alloy or steel with holes that create multiple attachment points. This can help eliminate the problem of tri-loading a connector.
- Pulleys – A pulley is a grooved wheel (called a "sheave") that is mounted on an axle or shaft. A rope may be placed into the groove and bent around the wheel to facilitate movement and change of direction, as shown in Figure 4-17. Pulleys are used in rope access for direction change, to create a mechanical

FIGURE 4-16
The use of rigging plates can help limit tri-loading carabiners. Courtesy of SMC – Seattle Manufacturing Corporation.

FIGURE 4-17
Pulley. Courtesy of Vertical Rescue Solutions by Pigeon Mountain Industries, Inc

advantage, and to transmit force. Only pulleys that are specifically designed for life safety use should be used to support a human load. Usually made of aluminum alloy, steel, or composites, pulleys should always be used within the limits of a specified rope diameter, manufacturer's strength ratings, and user-determined safety factors.

- Rope grabs – A rope grab is a device that, by means of squeezing or compressing a cam against a rope, moves easily along the rope in one direction but grips the rope tightly when pulled in the opposite direction. Different types of rope grabs are used for a variety of purposes in rope access applications, including as backup devices and ascending systems, as well as in rigging and hauling. Rope grabs intended for use as backup devices, and those designed for ascending, are discussed in Chapter 5. Rope grabs designed for more rigorous rigging use should meet higher performance criteria than personal ascenders. The camming mechanisms of rope grabs for rigging are designed with heavier loads in mind, and tend to spread the load more consistently across the rope.

The breaking strength of a rope grab is best determined by testing it on the rope upon which it is intended for use. For professional rescue applications where rope grabs are used as part of rigging, NFPA 1983-2012 specifies that a rope grab must hold 11 kN (2473 lbf) without permanent damage to the device or the rope it is tested on. This is one of the few existing standards for this type of rope grab. In the absence of specific criteria for rope grabs used for rigging beyond personal ascent/descent, it is wise to choose rope grabs that meet NFPA 1983 requirements (or equivalent) for rope grabs used in rigging, such as in haul systems or to hold a system in place,

- Braking Devices – Braking devices may be categorized as either personal descending or for lowering. Those intended for personal descent are typically called "descenders", and are discussed in Chapter 5. Some braking devices may also be appropriate for use in lowering and raising operations for a range of loads. In such cases they may be classified as rigging gear. There are two general classifications of braking devices: Simple friction and autolocking/bobbin style. No one braking device is perfect for every situation on the job, and choosing the right one requires some research, beginning with the type and diameter of rope used. Most bobbin-type and autolocking braking devices are intended for use with a specific range of rope diameters, whereas simple friction devices (like a brake rack or figure 8) are usually less restricted to a specific diameter (though generally speaking, a range of 10–12.5 mm is most common).

Consider the following when selecting a braking device for rigging:
  - Magnitude of the load for which the device is designed
  - Compatibility with rope and other equipment
  - Whether the device is more conducive for descending or lowering
  - Whether the device offers autolocking and/or panic-lock features
  - Whether the amount of friction may be adjusted during use for heavier loads or greater rope weight.

Many excellent braking devices are intended only for single person loads. Always verify that a device has sufficient friction to handle your intended load before putting it into a system.

4-11 MECHANICAL ANCHORAGE CONNECTORS

The term *mechanical anchorage connector* covers a wide variety of equipment options when applied to rope access and rigging. Anchorages are discussed in detail in Chapter 7, but here we will address several common types of anchorage connectors.

Basically, the term "anchorage" is most properly used to denote a place or fixture that supports and to which the various ropes and rope systems are attached, whereas an "anchorage connector" is a device or means of securing to an anchorage. In the case of a strap around a beam, the beam would be the anchorage and the strap would be the anchorage connector. If it is an eye-bolt in the beam, the beam would be the anchorage and the eye-bolt would be the anchorage connector.

Beam clamps, eye bolts, concrete or steel expansion/toggles, and portable high directional devices (generically referred to as gin poles, tripods, A frames, or quad-pods), are a few common types of mechanical anchorage connectors used by rope access technicians.

- *Portable high directionals* for the industrial rope access environment come in many shapes, options, and configurations. Their safe operation requires extensive training on resultant forces, critical angles, and the physics at play in their safe use. Portable high directionals are often set up as three-legged tripods, but depending on their adaptability, may also be rigged with four legs (called a quad-pod), only two legs (which is known as a bipod or A frame), or they may even use only one single leg (which is known as a monopod or gin pole). These variations are shown in Figure 4-18.

    The main benefit to the use of portable high directional is the elevation or redirection of the path of a rope, thereby reducing system friction or making for easier edge transitions. Use of portable high directionals is discussed in Chapter 7.

- *Beam clamps* are mechanical anchors that can be fastened to steel I beams to create an anchor point above a work area. Some are designed to be stationary,

FIGURE 4-18
Portable high directional anchors may be set up with four legs, three legs, two legs or a single leg. Courtesy of Pigeon Mountain Industries, Inc.

FIGURE 4-19
When using traveling beam clamps, make sure the flange to which they are attached is continuous. Courtesy of ClimbTech.

while others are designed to move horizontally with a technician. When using rolling or nonstationary beam clamps, as shown in Figure 4-19, care should be taken that there are no gaps or spaces cut into the flange on which the beam clamp travels, as this results in anchor failure when the clamp loses contact with the beam.

When loading beam clamps, it is important that they be loaded perpendicular to the flange of the steel I beam. If loaded at an angle, the beam clamp could be pulled sideways, thereby resulting in a drop of the load below or causing the beam clamp to slide off the end of a flange.

- The use of *expansion/toggle anchorage connectors* in both concrete and steel for rope access use are gaining popularity on jobsites, mainly due to improvements in technology and the increased availability of man-rated industrial options for anchoring. An example of these may be seen in Figure 4-20.

Whether permanent or removable, used in rock or steel, expansion/toggle anchorage connectors must be installed to the exacting specifications called out by the manufacturer in the user instructions. While expansion sleeves are used for installation in concrete or rock anchors, toggle locks are more common for use through a hole drilled in steel. They may be used individually, or in multiples to create a high strength redundant load sharing anchorage. Not all expansion/toggle anchorage connectors are rated for life safety, and therefore technicians need to ensure that they choose and use the correct anchorage connector for the job at hand (Figure 4-21).

4-12 ANCHOR SLINGS

Constructed of either synthetic material such as nylon or polyester or wire rope, *anchor slings* used for rope access should be capable of supporting the load appropriate for their intended use. When used for fall arrest and rope access, it is generally recommended that they be capable of supporting a life safety load of 5000 lbs (22 kN) per person. Commercially manufactured slings are typically used in one of two ways: A basket or choked configuration. These are shown in Figure 4-22. The basket configuration is usually the stronger of the two, but may have the tendency to slide up or down if it is not under load. Regardless of how they are configured, their MBS should not compromise the desired safety margin within the system.

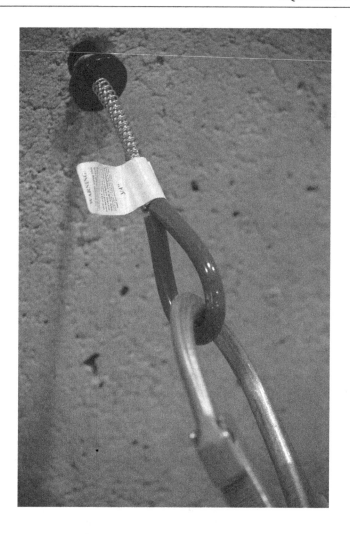

FIGURE 4-20
Removable expansion/toggle anchors rated for fall arrest and rope access use are gaining regulatory acceptance and popularity. Courtesy of Vertical Rescue Solutions by Pigeon Mountain Industries, Inc.

FIGURE 4-21
A high strength, redundant load sharing anchor built with removable expansion anchorage connectors. Courtesy of Tom Wood.

FIGURE 4-22
Anchor slings are usually used in a basket or choked configuration. Courtesy of Vertical Rescue Solutions by Pigeon Mountain Industries, Inc.

4-13 SUMMARY

When all is said and done, rope access technicians need to educate themselves about the equipment to which they entrust their lives. This education goes beyond usage in the field. It should include a thorough working knowledge of the proper ways to inspect, maintain, mark, store, and retire rope access equipment. Nearly all of this information can be found in the User Instructions that accompany the gear when it is purchased.

CHAPTER 5
Personal Equipment for Rope Access

By the end of this chapter, you should expect to understand:

- Examples of some items that may be classified as "personal gear"
- The difference between "personal gear" and PPE
- Essential requirements for life safety equipment
- How to choose a harness for rope access
- The purpose of a comfort seat
- How to select a helmet for rope access
- What a cow's tail is, and how to select one
- The purpose and function of a backup device
- Foundational requirements for connectors
- Some important considerations for descenders
- The difference between a handled ascender and a chest ascender.

5-1 INTRODUCTION

Equipment used in rope access can generally be separated into the categories of Personal Equipment and Rigging Equipment. Personal Equipment, which will be addressed in this chapter, typically refers to items that are worn or used directly by a worker. This includes (but is not limited to) equipment that is classified by regulatory guidance as Personal Protective Equipment (PPE). PPE is generally recognized as being equipment that is worn for the purpose of protecting a worker against certain recognized hazards. Examples of PPE might include helmets, harnesses, and gloves. However, the concept of Personal Equipment also extends beyond classic PPE to include certain items of access equipment, such as ascenders and descenders, which are specifically designed and intended for use by one technician at a time.

Professional Rope Access: A Guide To Working Safely at Height, First Edition. Loui McCurley.
© 2016 John Wiley & Sons, Inc. Published 2016 by John Wiley & Sons, Inc.

FIGURE 5-1
It is the responsibility of the employer to ensure equipment is suitable for the job. Courtesy of Vertical Rescue Solutions by Pigeon Mountain Industries, Inc.

Not all PPE used in rope access is necessarily specific to rope access. Depending on the task at hand, some examples of applicable PPE might include gloves, protective hearing protection (earplugs, muffs), specialized foot protection, eye protection, and respirators. PPE specific to environmental hazards is outside the scope of this text. The focus of this chapter is limited to PPE that is directly related to working safely at height.

Rigging equipment, addressed separately in Chapter 4, includes equipment that is used to create the working systems (the access system and backup system), such as ropes, anchorage connectors, connectors, and the like. For the purposes of this text, we will also discuss certain components of equipment that have a crossover function within the context of Rigging Equipment.

Responsibility for provision of PPE is a matter of some disagreement in some jurisdictions. In the United States, the Occupational Safety and Health Administration (OSHA) requires that employers protect employees from workplace hazards that can cause injury or illness. When applicable, this means that employers must provide PPE to users and ensure its use. Of course, for many rope access technicians, their equipment is personal and they prefer to own and use equipment of their own choosing. Under OSHA regulations this is certainly acceptable, but the use of PPE worker's already own must be completely voluntary – it cannot be mandated by the employer. Further, even when a worker provides his own PPE, it is the responsibility of the employer to ensure that the equipment is appropriate and adequate for the job. This may entail regular visual checks and verification of manufacturer requirements, as demonstrated in Figure 5-1.

5-2 ESSENTIAL REQUIREMENTS

All personal equipment that is used for life safety purposes, whether classified as PPE or not, should be designed and intended explicitly for use as life safety equipment, and should be selected based on their ability to perform adequately for the task at hand, as well as to meet applicable regulatory requirements for the jurisdiction in which the work will be performed. In some cases, the type of work being performed will dictate which regulation or standard applies for a given component of equipment. European standards, at least as of the writing of this text, do not take into consideration different work applications but instead address all types of a given piece of equipment under a single "Norm," or standard. Harnesses used for rope access, for example, will meet EN 891, the same standard as harnesses used for fall protection, or for rescue.

This differs from the standards in the United States, where PPE that is used for different purposes is regulated specifically according to its use. The regulatory authority that oversees workplace safety and health in the United States is OSHA. OSHA sets forth rules and standards that employers must meet to protect the worker, but specifications on product testing and performance requirements are largely absent from OSHA documents. Instead, these are deferred to national consensus standards. These national consensus standards are generally promulgated by specific industry organizations and are relative to that industry. As a result, there may be several different standards that could potentially apply to one type of product depending on where it is used and what it is used for. For example, in the case of the harness, described above, a harness for fall protection may be addressed by ANSI Z359 while a harness used for rescue might more appropriately meet NFPA 1983 or ASTM F1772.

Requirements for similar products may differ between standards for different industries, such as fall arrest, rescue, rope access, and ropes course work, because different types of work involve different hazards, different work methods, and different priorities. The advantage of using equipment that meets a standard specific to the type of use it will fulfill is that this is the only way you can know that the item meets the necessary requirements for that type of work. Again, following the case of the harness, ANSI Z359.12 mandates that a dorsal attachment be present for fall arrest, but does not mandate that a sternal or waist attachment be present for rope access. Here lies the problem: rope access cannot be performed safely without sternal and waist attachments. So, as you can see, a generic standard that mandates very specific requirements that fit one application very well might in fact reduce safety if used in another. When it comes to equipment, "safety" is largely dependent upon where, how, and in what capacity something (or someone) is employed.

As a starting point, all rope access equipment should be appropriate for the intended application. Because the rope access market is not (yet) large enough for rope access equipment to warrant regulation and standardization on a national level, it is imperative that the employer define the criteria required and specify this in the company's management system. When specifying equipment, both the type of protection to be used (rope access) and the conditions at the worksite should be taken into consideration.

Equipment that is used together in a rope access system need not necessarily come from one single manufacturer, but it is essential that purchasers ensure that components are compatible and that they function safely and effectively together. Compatibility is a subjective determination, and requires careful consideration by a competent person to ensure that the equipment that is used together is free from any characteristic(s) that would make it unsuitable for use with the other components to which it is joined.

In addition to compatibility, the employer's Competent Person should consider the anticipated loads within the system, specific to each device, to ensure that adequate safety margin will exist while in use (see Chapter 6). Of course, it should be assumed that equipment should only be used in accordance with manufacturer's instructions, and should not be modified or "repaired" without explicit permission from the employer to do so.

5-3 HARNESSES

In the early years of rope access it was common to see technicians employing seat/harness combinations that were separate from one another, or not integrally connected. These days, because many fall arrest standards and regulations mandate a full body harness, combined with the fact that many rope access technicians find themselves using rope access methods and conventional fall arrest or work positioning methods on the same work site, full body harnesses are a much more common fall arrest or work positioning – and arguably appropriate – choice.

Selecting a harness that meets existing standards for fall arrest is a reasonable starting place, but, at this writing, there is no one standard that, by itself, ensures adequacy for rope access. For this reason, the rope access user/purchaser is well advised to seek out the specific details prior to purchasing, and even to try the harness on, in a suspended position, to ensure its comfort and adequacy. Employers should refrain from purchasing equipment for rope access technicians without first obtaining guidance from a certified and competent rope access technician. Most fall arrest harnesses, as shown n Figure 5-2, do not have appropriate features for rope access, nor are they designed to suspend the worker comfortably in a sitting position from a waist or sternal attachment.

FIGURE 5-2
Fall arrest harness – Do not use for rope access. Courtesy of Reliance Industries, LLC.

HARNESSES

Harnesses selected for rope access, such as that shown in Figure 5-3, should be capable of supporting the wearer in a comfortable, upright working position when connected at the waist attachment. The comfort level when suspended from other attachments will vary, but this is something for which the user will quickly get a feel after trying a few samples. There should be sufficient adjustability in the harness to permit the users to vary their clothing and outerwear to accommodate different environments.

Many of the criteria for selecting a harness are objective; that is, they are specific and can be observed. These should be driven by the employer. Other criteria are subjective; that is, they are felt or perceived by the wearer. Subjective criteria are important, because wearer comfort and acceptance are integral to ensuring that the harness will be worn and used properly. Ideally, the person(s) who will wear the harness and work in it should have at least some input into harness selection with regard to at least the subjective criteria.

FIGURE 5-3
Suitable rope access harness.
©BEAL.

Selection considerations for harness

(a) Harness size, fit, and comfort
(b) Conformance with applicable regulatory requirements
(c) Supports the wearer ergonomically
(d) Appropriate location and dimensions of attachment points
(e) Compatibility with workseats, tool bags, and other auxiliary components
(f) Performance and adjustability of buckles
(g) Adequate material (webbing, thread, hardware) for the environment
(h) Bulk, if working in close spaces

5-4 A NOTE ABOUT COMFORT SEATS (SEATBOARDS)

Seatboards may be used as an adjunct to harness comfort. Properly used, these do not really justify classification as PPE, as their intent is solely for the purpose of providing comfort while suspended. The user is not fastened to the seatboard as they are into the harness, therefore it is the harness attachment that is more secure. For this reason, even when using a comfort seat, the technician's primary and secondary safety systems are attached directly to the harness rather than to the comfort seat. The seatboard (which may also be attached to the descender to facilitate adjustment) is only for comfort. Removal of the seatboard would not result in the loss of the technician's primary system attachment as well as in any loss of security.

One way to think of it is that if the descending/ascending system was attached to the seatboard, rather than to the harness, the user's safety could be compromised should they slip out of the seatboard. Conversely, with the descending/ascending system attached directly to the harness, the user's security is sure, even if they should come out of the seatboard.

For this reason, comfort seats are not regulated as PPE (unlike bosun's chairs, which are, at least in some industries, subject to specific standards). Selection of a comfort seat should be based on compatibility with other equipment, and user comfort (Figure 5-4).

FIGURE 5-4
Comfort seat designed to be used with rope access harness. Courtesy of Petzl.

5-5 HELMETS

All rope access technicians, whether performing work at height or serving on a ground crew, should wear appropriate protective headwear during rope access work operations. Helmets protect the wearer from falling objects, as well as protect the worker from obstructions in the unlikely event of a dynamic event, such as a pendulum or fall. Again, there is no one standard that can be relied upon to adequately encompass all the necessary performance specifications of a rope access helmet, so the employer must take some initiative here to specify what is appropriate for their application. Many rope access technicians choose helmets that are approved for mountaineering, by certified bodies such as by UIAA 106 (standard). This standard offers certain advantages over many common industrial hardhat standards, including side impact testing and greater chinstrap/retention system requirements. Choosing a helmet that meets both UIAA and an industrial helmet standard will help ensure safety and compliance. Helmet retention systems (chinstraps) should be designed to prevent the helmet from coming off the head, even in the event of a tumbling fall. This is best achieved through a three- or four-point retention system, to secure the helmet on the head and prevent it from shifting as shown in Figure 5-5.

In addition to these basic requirements, the employer should consider whether additional requirements are necessary for helmet performance with relevance to the type of work being performed; for example, adequate ventilation for work in hot environments, electrical protection for work in charged environments, or water drainage if falling into moving water is a hazard. Again, both the objective and subjective aspects of these criteria should be taken into consideration.

Selection considerations for helmets

(a) adequate protection for the type of work to be performed
(b) comfort (fit, weight and balance)
(c) adequate ventilation
(d) compatibility with other PPE (ear, eye, other protection)
(e) helmet does not restrict normal vision
(f) compatibility with communications equipment (if applicable)
(g) inclusion of three-point retention system

FIGURE 5-5
Rope access helmet with three-point retention. Courtesy of Pigeon Mountain Industries, Inc.

5-6 LANYARDS

Most lanyards are designated for a specific purpose: rope access, fall arrest, positioning, anchorage connection, and so on. Lanyards should be used only for the purpose that they are designed/intended for. The lanyards primarily used by rope access technicians to connect themselves to their system are commonly referred to as "Cow's Tails." A cow's tail lanyard differs from conventional lanyards in that it is generally made of dynamic climbing rope, rather than webbing. Dynamic climbing rope have inherent elongation properties that help absorb forces in the event of a fall, and is also appropriate for suspension during work (Figure 5-6).

At a minimum, cow's tail lanyards used for rope access should have a minimum breaking strength (MBS) rating of 5000 lbs (22.2 kN) and should be constructed from materials similar to the general performance characteristics of nylon. In some cases, cow's tail lanyards may need to be selected with consideration to the type of work being performed – with an eye toward heat or cut resistance, for example. Cow's tail lanyards made from steel cable or high-tech materials should be used only with great caution in rope access applications. These may have poor force absorption characteristics, poor heat resistance (UHMPE's), or be more susceptible to damage from repeated bending or other "normal" wear and tear (aramids). Common fiber characteristics as cited by the Cordage Institute are shown in Figure 5-7.

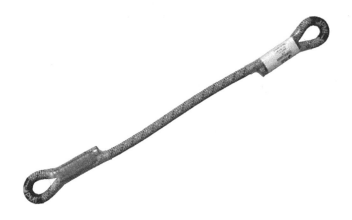

FIGURE 5-6
Cow's tail lanyard. Courtesy of Pigeon Mountain Industries, Inc.

FIGURE 5-7
Fiber characteristics. Data from the Cordage Institute.

	Fiber performance characteristics			
	Breaking tenacity (gpd)	Chemical susceptibility	Melting temp (°C)	Elongation at break (%)
Manila	5–6	Acids, alkalis	Chars @ 148	10–12
Sisal	4–5	Acids, seawater	Chars @ 148	10–12
Nylon	7.5–10.5	Mineral acids	218–258	15–28
Polyester	7–10	Sulfuric acids, alkalis	254–260	12–18
HMPE	25–44	Minimal	144–155	2.8–3.9
Para-Aramid	18–29	Strong acids and bases	Decomposes @ 500	1.5–4.4

Lanyards, as any equipment, should be selected with consideration toward how they fit together and perform with the entire subsystem.

> **Selection criteria for rope access cow's tails**
>
> - energy-absorbing characteristics
> - strength
> - compatibility with connectors (usually a screwlink)
> - suitability for work task or environment
> - adjustability, if desired
> - material
> - suitable length
> - appropriate terminations

While cow's tails with hand-tied terminations have historically been accepted and even desired, these have become less prevalent in recent years due to the difficulty in evaluating their quality and performance.

5-7 CONNECTORS

Hardware that connects parts of a system or subsystem together is considered to be a **connector**. The type of connector most commonly used for rope access is the carabiner. Carabiners are discussed in Chapter 5, Rigging Equipment, and as a rule it is generally appropriate to use rigging connectors for personal use as well. The same concepts already discussed in Chapter 4 also apply to connectors used for personal use.

Another type of connector known as a "screw-link", shown in Figure 5-8, is sometimes used as a personal connector in rope access. Screw links offer some distinct advantages over other types of connectors, primarily relative to geometry. For example, screw-links are less bulky than carabiners, thereby allowing more working space (e.g., at the harness attachment), they are less susceptible to cross-loading when attached to a single point, they hold chest ascenders and other equipment closer to the body, and they offer a high-strength connection. All screw links used in rope access work should be specifically manufactured to comply with life safety applications. Commodity screw links found at common hardware stores are not appropriate for life safety use.

Screw links are most appropriately used in locations where their unique dimensions offer additional security, such as for connecting a cow's tail or chest ascender to a harness, or even as a harness buckle. One concern with screw links is that they do not have a self-closing, self-locking gate, so they may be more susceptible to being left open. For this reason, screw links should preferentially be used as an integral connection. An integral connection is one that must be made with, and cannot be removed without, the use of a tool. As such, screw links should be tightened with a tool when placed into service, and not removed until the work is completed. This will help to ensure that they are closed properly before use.

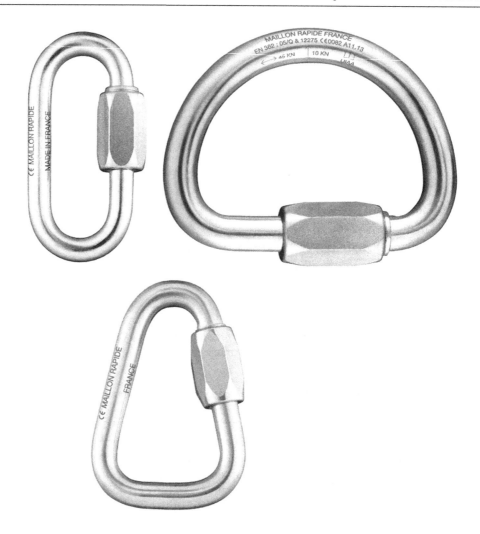

FIGURE 5-8
Some examples of screw links.
Courtesy of EDELWEISS SAS.

> Screw links are sometimes used as a harness buckle. This is an acceptable use of a screw link, as long as it is permitted by applicable regulation, and as long as the manufacturer has designed the harness with this in mind.

5-8 DESCENDING DEVICES

A descending device (also known as "descender") is a type of braking device used to create friction between the rope access technician and the rope for the purpose of controlling the movement of the technician along the rope, as shown in Figure 5-9. The descender is typically connected to the technician by means of a connector between it and the harness' waist attachment point. The rope is reeved through the descender as per the manufacturer's instructions, and the descender is used accordingly. Descenders in which the rope is reeved through the descender in a linear

DESCENDING DEVICES

FIGURE 5-9
Descender in use. Courtesy of Pigeon Mountain Industries, Inc.

fashion, rather than requiring the rope to wrap or twist, help to keep the rope from becoming irreparably twisted and tangled.

Most descending devices rely on friction to perform their function, so the technician must be aware of the potential for heat buildup – which is a natural result of friction. Heat buildup is increased by heavier weights and/or faster descents. In extreme cases, heat buildup in descenders can damage rope or other equipment, cause second degree burns to bare skin, and have other adverse effects. For best results, choose a descender that dissipates heat, and avoid descending too quickly to help control heat buildup in the device you are using.

A descender must be of a design that is compatible with other components of the system – especially the host rope. This concept of compatibility includes a variety of factors including how well the descender grips the host rope. If the grip is too aggressive, the descender can be difficult to operate or even cause damage to the rope. On the other hand, if it is not aggressive enough, it may not offer sufficient friction to control the load.

Most descenders are intended for use with a wide range of ropes within specified diameters. Even braking devices that are not designated by the manufacturer as being rope-specific will vary in performance from rope to rope depending on everything from rope diameter to construction to material to the mass of the load and the distance of travel. It is essential that the competent person be capable of making an educated decision regarding the compatibility and use of braking devices for a given circumstance based on a combination of product instructions, education, and experience.

There are a number of optional features that a user might prefer to choose in a descender, depending upon the intended use. Devices that automatically stop when the user lets go are known as "Auto-locking devices." These are often prescribed by jurisdictional authorities, especially for technicians whose experience is limited, for the added security they offer. Auto-locking descenders, as shown in Figure 5-10 often feature a handle that must be levered to allow the technician to descend, while the technician's other hand (known as the brake hand) grips the rope to control speed.

In addition to requiring both hands to be occupied, another disadvantage to autolocking descenders is that in the case of long descents the weight of the rope below can cause the auto-locking mechanism to engage, thereby preventing descent. For long descents, or where the mass of the load may vary significantly, choosing a descender with adjustable friction such as a brake rack, shown in Figure 5-11, adds

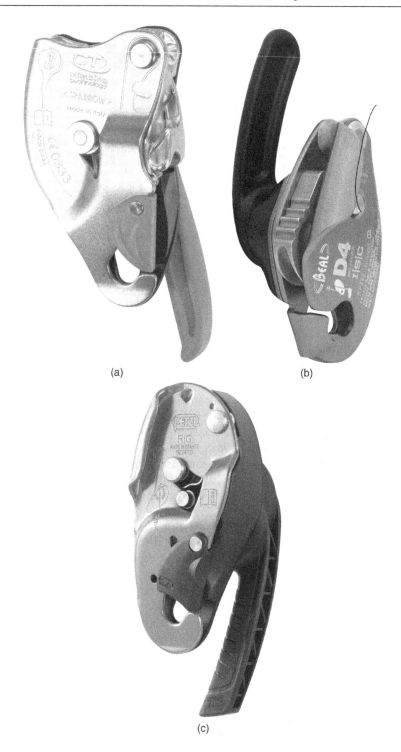

FIGURE 5-10
Several types of autolocking descenders. a) Sparrow, Courtesy of Climbing Technologies, b) ISC D4, Courtesy of Beal, c) Rig, Courtesy of Petzl.

a degree of safety. A brake rack may be used either as a descender or as a lowering device. Although a brake rack does not feature auto-locking capabilities, it dissipates heat well and the ability to add and remove bars (friction) makes it a better choice for very long drops, as for example, at heights above 300 feet. The ability to adjust friction is also useful in other situations, such as where the weight of the load might change during the course of the work. In this the Competent Person must choose

FIGURE 5-11
Brake rack. Courtesy of Trask Bradbury.

wisely as there are, at least at this writing, no auto-locking devices on the market that also offer adjustable friction.

One alternative means of mitigating the undesirable friction from rope weight when using an auto-locking descender for a long drop is for the technician to carry the "tail" end of the rope in a rope bag that is attached to themselves or their descender. In this operational "work-around" the rope can be paid out from the bag as needed during descent rather than hanging below where it could impart excessive tension on the braking end of the rope. Use of a short rebelay is an acceptable method of rope weight mitigation as well.

Regardless of whether or not a device features an auto-locking mechanism, the users should take precautions to prevent accidental activation or descent if at any time they intend to remain stationary on the rope for a period of time or need to release their brake hand for some reason. Known colloquially as "locking off," one common means of accomplishing this is to push a bight of rope through the connector to which the descender is attached, then loop that bight up and over the top of the device as shown in Figure 5-12. Other, device-specific methods for locking off may be prescribed by the manufacturer.

Another feature offered by some descenders is that of a panic lock. This term refers to any mechanism that causes the descent to halt if the technician grips too tightly. Found primarily in auto-locking devices that require activation of a handle

FIGURE 5-12
A generic method of locking off. Vertical Rescue Solutions by Pigeon Mountain Industries, Inc.

for travel, the panic lock protects against the natural reaction of a person to tense up and grab that handle in panic when in fear.

Before using a descender at height, it is advisable to practice using it on a level or gently sloping surface in a controlled environment to ensure familiarity with its operation before committing to it for life safety.

> **Selection criteria for descending devices**
>
> - Panic lock, if desired
> - Hands free auto-lock, if desired
> - heat dissipation
> - mass of intended load
> - length of descent
> - effect on rope (wear or twist)
> - compatibility with rope type and diameter
> - ease of use

Some descenders may also be used as belay devices. A belay is a method used by one technician (belayer) to provide protection against a fall for another technician (climber), as shown in Figure 5-13

The use of such systems, which is discussed in greater detail in Chapter 11, requires special skill and training. A belay device should permit the belayer to feed rope through in either direction at sufficient speed to allow the belayed person to move at a comfortable rate of speed, and should allow the belayer to catch a fall without undue force.

5-9 ROPE ACCESS BACKUP DEVICES

A rope access backup device is a type of rope grab that is designed to be used as part of the technician's secondary system, to stop a fall in the unlikely event that

ROPE ACCESS BACKUP DEVICES

FIGURE 5-13
Descender used as belay device.

the primary system would fail. The backup device is attached to the safety line via the technician's cow's tail (or sometimes a manufacturer's shock absorber), and it functions by gripping the safety rope in the event of a fall or mainline failure.

The rope access backup device performs a function similar to that of a fall arrester in a conventional fall arrest system, but accommodates the unique needs of rope access that must also be considered. The backup device should travel up the rope easily, perhaps even trailing, as the user moves upward, but if the person falls the backup device will engage gripping the rope with enough force to stop the fall without damaging the rope. A certain amount of "slip" may be designed into the backup device for this purpose.

A variety of backup devices are available. Some of these are shown in Figure 5-14. No one device is ideal for all circumstances, but each offers a unique set of advantages and disadvantages. The "best" choice for any given situation depends on such factors as technician experience, work type, rope selection, direction of movement, and anticipated load.

Due to the unique nature of rope access work, backup devices used in this field must be relatively easy to put on and take off a rope and must be of a design that can be installed onto a horizontal or oddly positioned rope so as to facilitate long rebelays, rope-to-rope transfers, and other transitions, yet still not be susceptible to inadvertent removal. Further, the user must consider potential for a two person load,

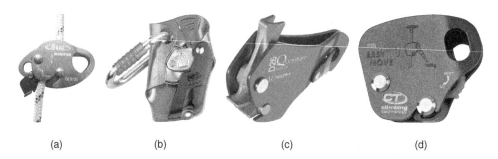

FIGURE 5-14 a) Monitor, courtesy of Beal, b) ASAP courtesy of Petzl, c) heightec Vector, courtesy of PMI, d) Easy Move, courtesy of Climbing Technologies.

such as during rescue. For these reasons, devices that are certified to typical "fall arrest" standards may not necessarily be appropriate for use in rope access. Backup devices should meet applicable requirements of the country/jurisdiction where the work is taking place.

Selection criteria for backup devices

- energy-absorbing capabilities
- ability to self-manage (e.g., needing little or no user intervention)
- ability to arrest a fall gradually rather than suddenly
- compatibility with rope type and diameter
- suitability for work environment, e.g., ice on the rope
- suitability for body weight and work method being undertaken
- ease of unloading after a catch
- ability to position the device on the rope
- compatibility with a rescue load
- ability to install on a horizontal or oddly positioned rope

5-10 ASCENDERS

Rope grabs specifically designed for gripping a rope for the purpose of ascending it are known as Personal Ascenders. Some of the unique characteristics of ascenders include a handle for gripping, a more aggressive camming device, and a design that makes the unit easy to place on and take off the rope. Models that feature a handle are called "handled ascenders" (Figure 5-15) while "chest ascenders" (Figure 5-16) feature connection points designed for rigging into the sternal part of a harness and a slight twist in the frame to facilitate better alignment when rigged.

Other configurations exist, but these are the most common. Personal Ascenders are intended for ascending with only one person's body weight, not for hauling systems in which forces are multiplied. Rope grabs for rigging are addressed in Chapter 5. Personal Ascenders are usually not designed to withstand a fall, and are not recommended as a primary safety attachment. Especially those with aggressively toothed cams tend to be very unforgiving in the event of misuse, so Personal Ascenders should be used with great care and excessive loading must be avoided.

ASCENDERS

87

FIGURE 5-15
Handled ascender. Courtesy of Pigeon Mountain Industries, Inc.

FIGURE 5-16
Chest ascender. Courtesy of Pigeon Mountain Industries, Inc.

> **Selection criteria for ascenders**
>
> - ease of use
> - security on rope
> - compatibility with rope and other components
> - functional performance
> - ability to grip
> - right hand vs left hand orientation

5-11 GLOVES

Gloves are useful equipment for a rope access technician to use in protecting hands against abrasion and from frictional heat when moving along a rope. Gloves may be of leather or a suitable synthetic, with extra layers to protect the hand from heat and abrasion from a running rope. Finger dexterity is useful for rigging, while extra layers of protection through the palm and the groove between the thumb and fingers is desirable when handling moving ropes, such as through a brake device. Palms of the glove should be double, or even triple-layer leather or leather-like material to allow the hand to slide easily along the rope yet prevent transference of heat to the hand. Some technicians are "palm grippers" and for these the protection across the palm is paramount. For thumb pinchers, however, the extra layers of protective leather should extend between the thumb and forefinger, toward the back of the hand, as shown in Figure 5-17.

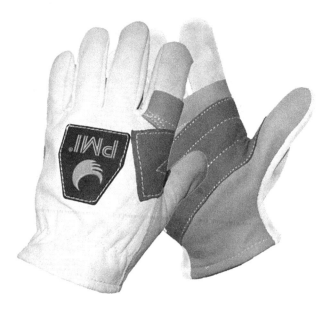

FIGURE 5-17
A glove designed for thumb pinchers. Courtesy of Pigeon Mountain Industries, Inc.

Gloves should be selected to accommodate the ergonomic needs of the technician, including finger dexterity and location of protection, and to provide adequate protection for the task. Some technicians prefer to use fingerless gloves, which work well for handling rope but may not protect the fingers from other hazards. Where a variety of hazards exist, priorities must be carefully balanced. For example, a glove designed for rope work may not provide sufficient protection against other hazards, such as flame or blood borne pathogens, while a cut resistant glove or a glove designed to withstand flame, may not provide sufficient dexterity for rope work. In such a case, the users may need to carry both, or may need to prioritize and base their choice on the comparative levels of potential hazards.

5-12 CLOTHING AND PERSONAL WEAR

For most jobs, rope access technicians may be able to wear normal work attire that is appropriate to the worksite and task. Clothing should provide appropriate coverage, be in good condition, and free from rips, tears, or ragged edges to help prevent it being snagged or caught in equipment or machinery. Similarly, baggy clothing and loose shirttails can present a hazard for snagging. Clothing intended to protect the technician from job-specific hazards, such as heat, sparks, flame, weather, and/or wet conditions, should be selected based on the conditions in which the employee is working on a given day.

Footwear for rope access technicians should be sturdy, with stiff soles and durable uppers to provide protection from ladder rungs, hard surfaces, and dropped objects. Special attention should be given to nonslip soles, ankle protection, and foot support.

5-13 OTHER PPE

Additional PPE may be required for the safety of the technician, depending on the assigned task and the environment. Examples may include common items such as safety glasses, sunglasses, and hearing protection, or more specialized equipment such as breathing apparatus, personal flotation devices, and flame-resistant clothing. Equipment should fit properly, be compatible with other items of equipment being used and should not create undue hazards or hindrance to the worker. It is the responsibility of the employer to ensure that the worker is protected against foreseeable risks to health and safety.

5-14 EQUIPMENT TRACEABILITY AND RECORDKEEPING

All load-bearing rope access equipment should be indelibly marked with a unique identifier and records should be kept on that item through its entire lifespan. For any given item, records should indicate what relevant certifications or regulatory requirements are met by that item, as well as when the item was placed in service and the results of inspections.

The following information should be maintained in product records:

(a) unique identifier for that item
(b) date of manufacture
(c) date item was placed in service
(d) information regarding significant/extraordinary use
(e) storage location
(f) details of repairs or modifications
(g) dates, type, and results of inspections

5-15 SUMMARY

PPE includes those items of safety equipment that are worn or used directly by a worker for the purpose of protecting them against certain recognized hazards. In addition to the specific rope access equipment addressed by this chapter, employers should consider whether additional safety equipment is necessary, based on the worksite and job hazards. Both objective and subjective criteria should be taken into consideration when selecting PPE. PPE should be properly cared for, in accordance with manufacturer's instructions, and once placed in service should be subject to ongoing maintenance and inspection requirements as established by the employer.

Additional information regarding inspection and care of equipment may be found in Chapter 22.

SECTION 2

Skills for the Rope Access Technician

CHAPTER 6
Rigging Concepts

By the end of this chapter, you should expect to understand:

- The effects of gravity on a rope access system
- How the fall line affects rigging
- The effects of friction on a rope access system
- The methods for protecting against the effects of friction
- Forces in a rope access system
- What a vector is
- The effect of angles in a system
- How to estimate angles
- How multiple simultaneous vectors influence a system
- What Mechanical Advantage (MA) is, and how to create it
- The difference between Actual Mechanical Advantage (AMA) and Theoretical Mechanical Advantage (TMA)
- How to determine TMA
- What a load ratio is
- What a safety factor is used for in rope access
- How to calculate the safety factor of a system.

This chapter will be useful in laying the groundwork for understanding the skills and methods described throughout the remainder of this text. You will not find specific information on how to perform tasks in this chapter, but you will find the tools necessary for understanding the fundamental ideas and characteristics behind what makes some of the techniques that are discussed later work more effectively than the others. Understanding the concepts in this chapter will help you better apply the specific techniques that will be discussed throughout the remainder of this text.

The effectiveness of rope access equipment, skills, and techniques is dependent on two key principles: gravity and friction.

Professional Rope Access: A Guide To Working Safely at Height, First Edition. Loui McCurley.
© 2016 John Wiley & Sons, Inc. Published 2016 by John Wiley & Sons, Inc.

6-1 PRINCIPLES AT WORK IN A SYSTEM

Gravity

When we speak of gravity, we are simply speaking of the attraction that two masses have for one another. For our purpose the attraction is between the earth and objects on or near its surface. On earth, gravitational attraction can be generally quantified using an approximate gravitational attraction of 9.81 m/s². Although the precise calculation varies on different parts of the globe, earth's gravity is more or less constant. It is only when we learn to pay attention to, and adequately predict the effects of gravity, that it becomes a useful tool for our rigging toolbox.

Gravity tends to pull straight down, toward the earth's center of mass. Therefore, any load – including a technician – suspended by rope and left to hang freely can similarly be expected to follow this natural gravitational pull. Like a plumb-bob, in the absence of any interference the suspended technician will dangle straight down from his anchorage toward the earth. By definition, the amount of force with which the earth attracts is known as the weight of the technician. The force of gravity near the surface of the earth is equal to the weight of the object, as found by the equation:

$$F_{grav} = m * g$$

where g = the gravitational constant and m = mass (in kg)

Friction

The next most common influence on our rigging, friction, can be described simply as the resistance of an object against the surface on or through which it is trying to move. The effect of friction is most often realized in opposition to the motion of an object. For example, if you push a brick across the floor, there will be friction between the bottom of the brick and the floor. Just how much friction there is will depend upon the characteristics of the contact surfaces (known as the "co-efficient of friction") and how firmly they are pressed together. This concept is expressed by the equation:

$$F_r = \mu * P$$

where, F_r = resistive force of friction; μ = coefficient of friction between the two surfaces; P = Perpendicular force pressing the two surfaces together.

In rope access, friction occurs between the technician and the surfaces with which one comes in contact, between the rope and other equipment, and between the rope and various environmental factors such as terrain. When planned into the system and used wisely, friction can help the technician maintain control. When not planned for, friction can create hazards, damage equipment, and jeopardize the work. Friction is generally thought of as being "good" during lowering events and "bad" during raising, but as we will see there are always exceptions.

Perhaps the most common and least controlled friction mechanisms in rope access are the edges over which the rope bends from a near horizontal to a near vertical orientation when a load is suspended, as shown in Figure 6-1. Such friction may be imparted once or more, or not at all, in a given system, depending on the terrain. This type of friction not only affects the efficiency of the system's performance, but can also have potentially damaging effects on the equipment.

FIGURE 6-1
Contact with an edge imparts friction on a system. Courtesy of Abseilon USA, www.abseilon.com.

FIGURE 6-2
A technician redirected by contact with a sloped surface. Courtesy of Abseilon USA, www.abseilon.com.

One common example of friction in a rope access system might be the contact between the technician and the surface beneath their boot. The direction of the technician, and the direction of the forces acting on the system, will be redirected based on the effects of that surface, as illustrated in Figure 6-2.

The significance of this redirection will be determined by the relationship between the technician's boot and the surface upon which it has contact, as well as the weight of the technician. Still, gravity, as offset by this surface contact, will pull the technician in the path of least resistance toward the earth, with the path altered primarily by terrain features and friction. This concept is commonly described as the "fall line." The term fall line is used to describe the resulting line of travel of a suspended object as it is redirected and influenced by surfaces and friction within

FIGURE 6-3
Canvas edge pad. Courtesy of PMI.

the system. Simply described, it is essentially *the natural path a ball would take if rolled down a slope*.

With the quest for anti-gravity methods still elusive to the common man, it is easier to learn to use this relationship between gravity and friction to our advantage than to fight it. Understanding the relationship between gravity and friction, and the resulting fall line, is fundamental to effective rigging in rope access. Understanding these will help the technician better control the path and rate of movement.

Anything with which a rope or other soft goods comes into frictional contact has the potential to cause damage. In general, a sharp edge is generally more of a concern than a smooth one, and an extreme bend more potentially damaging than a gentle one. The amount of weight suspended by the rope, the length of rope that comes into contact with the edge, and the motion of the rope in use, will all influence the level of potential damage.

Modern kernmantle life safety ropes offer excellent protection from frictional damage, not just because of the toughness of the braid but also because the core-and-sheath design protects the primary load-bearing core from being subjected to damage. Additional protection can be achieved through the use of a variety of types of protective equipment such as edge pads (Figure 6-3) and edge rollers (Figure 6-4).

Either edge pads or rollers may be used for protecting stationary ropes, but edge rollers are often the better choice for moving ropes. Moving ropes are likely to damage textile rope pads, rendering them ineffective. Low-friction edge pads made of plastic (Figure 6-5) or other smooth materials, though not quite as efficient as rollers, may also be used for moving ropes.

The savvy technician will plan to use friction to their advantage in a system. It is through friction that we control the rate and direction in which a load moves, and there are certain components of equipment that are specifically designed for this purpose.

Friction is an important part of how descenders, ascenders, and other rope adjusters offer control. Aside from friction against the surface(s) of the device, control may also be achieved through the "pinching" of the rope between two surfaces, and/or the angles made by the rope as it is reeved through the device.

Angles

Angles occur wherever a rope or other tension member changes direction in a system. Transition points where a rope goes over an edge creates an angle, as does

FIGURE 6-4
Edge roller. Courtesy of Trask Bradbury.

nearly every point where a piece of auxiliary equipment (such as a pulley, connector, etc.) is used in conjunction with the rope (Attaway, 1999).[1] Angles are also an inevitable part of the knots and bends used to terminate a rope.

Angles are a consideration throughout the rope access system, including anchor rigging, vertical lines, horizontal lines; practically anywhere rope or webbing is used. As in the case of friction, properly managed angles are a useful part of rope access, while unplanned-for angles can be detrimental to the system. The two primary considerations when dealing with angles are strength loss of material, and multiplication of forces.

Wherever a rope incurs a bend, it is prudent to assume at least some loss in strength of the rope at that point. How much loss depends on the nature of the bend, and whether it is around a tight or a narrow pivot point. This concept is typically described in terms of the "d:D ratio", which is illustrated in Figure 6-6. Where the bend in the rope is tighter than 1× the diameter of the rope, the d:D ratio is said to be less than 1. Strength loss of a rope around a bend is generally said to be evident at a d:D ratio of 2 (i.e., the bend in the rope is approximately 2× its own diameter), and begins to become significant at a d:D ratio of 1. This is not to say that a d:D ratio of less than 1 should be avoided – this is impractical. It is only to say that the technician should be aware of this reduction in strength, and should take this strength loss into consideration when calculating forces and system safety factors.

When a bend occurs in a fixed (stationary) rope, the primary consideration is strength. When as part of a moving rope system, such as over an edge or pulley (block), energy loss due to the travel of the rope around the bend also becomes a consideration. This concept of energy loss is referred to as efficiency.

A system wherein a rope travels in a straight line without contacting any object encounters no efficiency loss, whereas in a pulley system that involves the bending of a rope around a sheave there will be some efficiency loss, even if the pulley

---

[1]The Mechanics of Friction in Rope Rescue. Stephen W. Attaway, Ph.D. (attaway@highfiber.com). International Technical Rescue Symposium (ITRS 99) Proceedings.

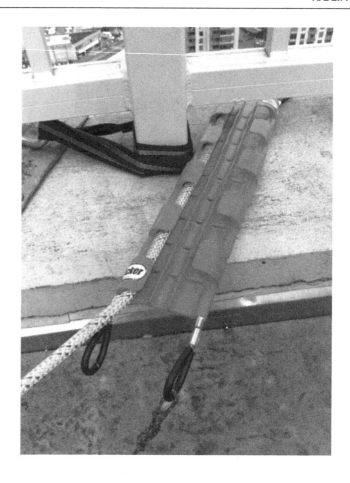

FIGURE 6-5
Low friction edge pad. Courtesy of SMC – Seattle Manufacturing Corporation.

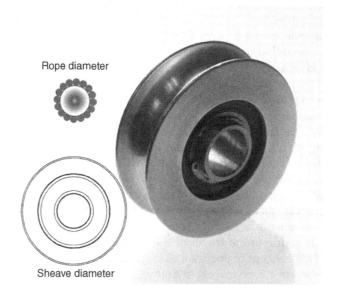

FIGURE 6-6
d:D ratio.

PRINCIPLES AT WORK IN A SYSTEM

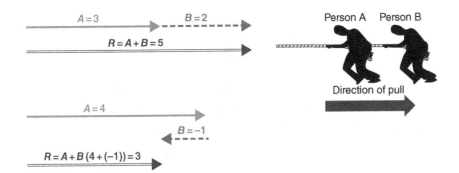

FIGURE 6-7
Vector addition. Courtesy of Robert McCurley.

> The tip-to-tail method simply represents the force exerted on something with vectors that indicate the magnitude of force and the direction of the force. When these forces are in a straight line, the magnitude of the resulting force is simply a matter of adding the individual forces together. By placing the tail of each successive vector at the tip of the preceding vector, the vectors are simply added to get the resulting force with the direction of force shown by drawing the resultant vector from the tail of the first vector to the tip of the final vector to be added.

has very efficient ball bearings. As in the case of strength loss, the tighter the bend the greater the efficiency loss. Here, however, we tend to speak in terms of radius rather than diameter. All other things being equal, engineering principles estimate that efficiency loss as a result of a bend in a moving rope system becomes significant when the bend is less than four times the radius of the rope.

The other factor to consider when dealing with angles is the possibility that the angle is causing an increase in force applied to the system. Force magnification occurs when multiple forces simultaneously act upon a point.

Vector Forces

Colloquially, rope technicians often use the term "Vector Force" to describe the concept of multidirectional vectors that occur as a result of, or as a part of, an angle. However, this usage is not accurate. The term "vector" actually simply describes a directional force.

There are two components to a vector: direction and magnitude. For example, the force of gravity pulling down on the technician creates a vector, or directional force. This vector must be offset by an equal and opposite force – for our purposes, the tension in a rope – if the technician is to remain stationary.

The net force exerted on an object is the sum (also known as the resultant) of all of the force vectors acting upon it. Vector addition is fairly straightforward when all of the forces are acting in one linear direction. This concept is illustrated for a straight line system by the equations in Figure 6-7.

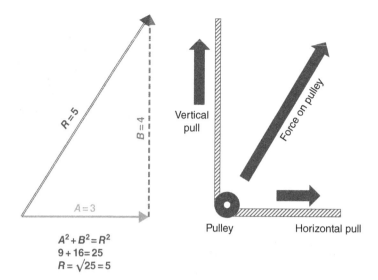

FIGURE 6-8
Tip-to-tail addition of vectors. Courtesy of Robert McCurley.

When forces are equal and opposed, a load remains stationary. That is, the load does not move. When the forces are imbalanced (disregarding friction), acceleration occurs and the load will move in the direction of the higher force. The rate and direction of motion that does (or does not) occur in a system is a direct result of the magnitude of the forces. Friction, discussed earlier in this chapter, is often a key factor in creating resistance to oppose the downward motion of a load.

Where multiple vectors act simultaneously on a load from different directions, the various forces that are applied to that point at the same time will affect the characteristics and performance of the system. In rope access this can occur, for example, when a pulley is placed into a system or when the rope travels over an edge. Multidirectional force vectors act on the load in multiple directions – thereby changing how force is applied on anchors and other system components. Here, calculation of the force vectors acting on the singular point can be a little more complex.

The magnitude of pull in each of these directions and the resulting forces in the system can be quantified using simple tip-to-tail addition methods, as illustrated in Figure 6-8. This illustration shows how multiple vectors exerted from different directions on a given point can result in a direction and magnitude of force that differs from the original vector(s). While these calculations can become quite complex as more vectors are added, the key takeaway for the technician is to realize that multidirectional vectors can affect a system well beyond the concept of simple opposing forces. The sum of the forces added together gives what is commonly called a "resultant." This term actually refers to what is more accurately called a resultant vector force.

> The tip-to-tail method also works for forces that are not in a straight line. For example, if a pulley experiences two forces that are exerted in directions that are perpendicular to one another, the resulting magnitude and direction of force can be determined by adding the vectors representing the forces using the tip-to-tail method.

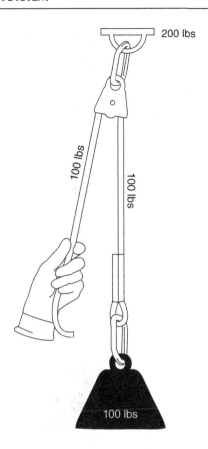

FIGURE 6-9
Suspended load, supported by 2 parallel, equally loaded legs of a rope around a bollard. Illustration by Margaret Deluca.

The concept of vectors can sometimes make more sense within the context of an actual system. As an example, consider that if a load weighs 100 kg and is equally supported by two parallel "legs" of a single rope, as illustrated in Figure 6-9, each line supports only about half the combined load and the force on the anchor is no greater than the force of the load itself. In this case, each line would support approximately 50 kg and the combined force at the vectored line on the anchor would be 100 kg.

If we rigged only systems wherein the rope was bent around a bollard then connected via both ends of that same rope to the load, there would be little more to say about vector force multiplication in this text. However, this is not the case in typical rope access systems. Lines are rigged over edges, around anchors, and through various components of auxiliary equipment, and all of these variations affect the directional forces.

Again assuming that the load weighs 100 kg, but this time assuming that it is connected to only one end of the rope, in order to prevent the load from falling, an equal force (100 kg) must be exerted on the other leg, as illustrated in Figure 6-10. With relatively parallel legs, the force at the anchor is equivalent to the combined load on the two legs — that is, 2× the load, or 200 kg force. With the forces being equal, the load remains stationary, but note that the force at the anchor is now twice that of the load itself.

If the intent is to move the load, additional force must be exerted on the other end of the line to initiate the movement of the load. Assuming the aforementioned 100 kg load, any amount in excess of 100 kg applied to the other end of the rope will begin to move the load (system efficiency notwithstanding). The force at the

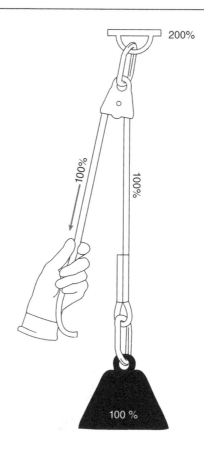

FIGURE 6-10
Suspended stationary load, 1 line attached, opposite end held by hand. Illustration by Margaret Deluca.

anchor will still be roughly the equivalent of the combined load on the two legs. Assuming, for example, that we are exerting 100 kg to hold the load steady, but then add another 10 kg force in order to move the load, (Figure 6-11). In this case, the force at the anchor will be the equivalent of the sum of all the parts: 100 kg load + 100 kg holding force + 10 kg force to move the load, or about 210 kg.[2] As you can see, the consequence of this is again increased force at the anchor.

Note that there is no MA provided by the pulley in this system. It serves merely as a change of direction. The force required to move the load must be at least equal to the force exerted by the weight of the load itself (plus acceleration). It can be said that there is a 1:1 relationship between these forces.

Now, let's flip this illustration upside down, placing the pulley on the 100 kg load rather than at the anchor.

Of course, the inverse also holds true. In Figure 6-12, with the pulley at the load, the weight of the load is equally shared between the two lines – one to the anchor and the other to the technician – with each experiencing only 50% of the total weight of the load (plus a bit for acceleration, if applicable). As a result, the anchor sees only half the weight of the load. This is an example of how a vector force can be adapted to reduce system forces.

---

[2]The examples given throughout this segment of the text are approximate, for illustration purposes, and do not take into account efficiency or friction.

PRINCIPLES AT WORK IN A SYSTEM 103

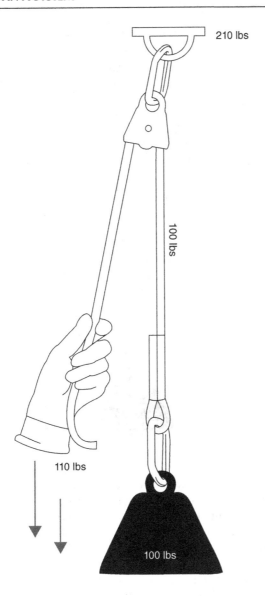

FIGURE 6-11
Anchor forces increase as enough force is applied to move the load. Illustration by Margaret Deluca.

Also notable in this case is that the force exerted at the working end of the rope is effectively doubled by the influence of the pulley and the effect of the anchor providing half the support. There is, in effect, a 2:1 MA provided by this system.

This is the premise that provides the foundation for MA systems that rope access technicians use to move loads and perform rescue. Generally speaking, with exceptions that are beyond the scope of this text, pulleys at the load (or "moving") end of such a system provideMA, while pulleys at the anchor (or "stationary") end of such a system result only in change of direction.

Remember that the term "vector" simply refers to the magnitude and direction of a force. As you can see, where the effects of a force are shared across multiple directions, the result can either increase or decrease the magnitude of the force in

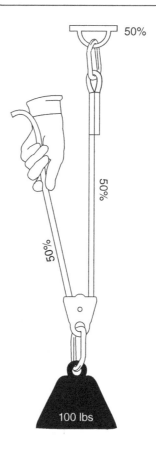

FIGURE 6-12
Pulley at load rather than anchor, one end anchored. Illustration by Margaret Deluca.

each of the directions. The ability of the technician to effectively monitor system safety is dependent on their understanding this concept. This is important not only in pulley-based systems, but also in stationary systems such as anchors. In a stationary system, where a line is anchored at each end and a direction change is introduced midline, the combined total of forces at each of the anchored ends can be significantly higher than that of the load.

Try this experiment: have two coworkers stand 10 ft apart holding a rope taut between them. Then press down on the rope mid-span. What happens? The people holding the ends of the rope cannot maintain the tension! Inevitably, the introduction of a directional force – or a vector – on the middle of the rope increases the force exerted at the ends.

The key point here is that as the angle increases, so does the resulting vector – that is, the magnitude of directional force – exerted on the anchors.

On a practical level, these concepts come into play when rigging anchors, highlines, and hauling systems, among other things. To effectively manage the system, the technician must understand how vectors play into each part of their system, and how these change as the system moves. Force magnification becomes even more significant when the concepts of acceleration and vectoring are combined.

It is not essential for the average technician to be able to calculate this to the nth degree, but as may be apparent by now these are important concepts to at least

PRINCIPLES AT WORK IN A SYSTEM

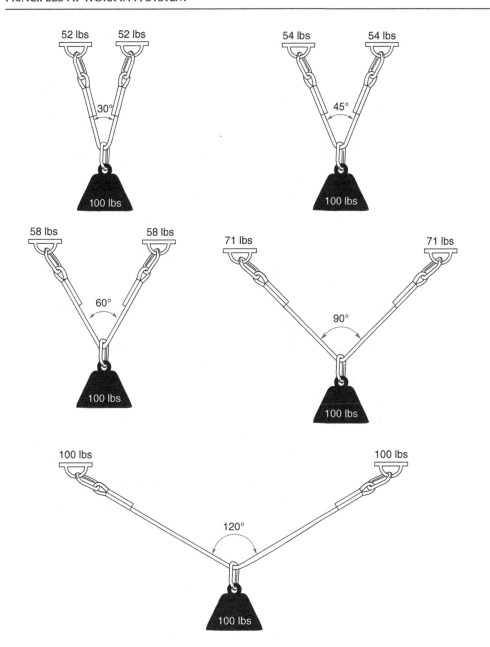

FIGURE 6-13
Increase in angles increases force at anchors. Illustration by Margaret Deluca.

recognize and understand when rigging. It is important to understand how an angle influences a vector (force) both at the respective ends of the rope(s) and at the point of the angle. The increase in force is seen not at the point of the vector, or angle, but at the extensions of the lines that feed in to the angle, as shown in Figure 6-13. At the point of the vector, the force decreases relative to the extensions as the interior angle increases.

The magnitude of a force as it is magnified by the imposition of an angle is the concept to which rope access technicians are most often referring to when they

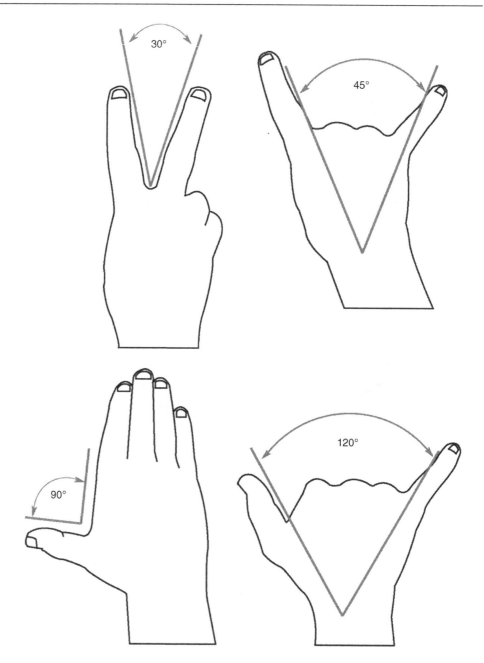

FIGURE 6-14
Estimating angles using your hand. Illustration by Margaret DeLuca.

speak of a vector force. This diagram is a useful one to memorize, or at least a few key angles. As a mnemonic, the technician can use their own hand as a simple tool to estimate angles in their system, as shown in Figure 6-14. Of course, hand calibration is not particularly precise, so this method should be used only for approximation, and with caution.

Understanding the significance of vector forces is essential for every rope access technician. The concepts are applicable to the simplest ascending and descending systems and to the most complex directional deviations and haul systems. We will explore these ideas in relation to specific methods and systems in the chapters to come.

6-2 USING THE PRINCIPLES

Mechanical Advantage

No discussion of forces would be complete without some attention given to how these concepts relate to MA.

When it becomes necessary to raise a load, the most direct method is to simply pull on the end of the rope. In this case, there would be no MA. As noted above, there would be a 1:1 relationship between the load and the force required to suspend it. To move the load, an even greater force would be required. In other words, a force greater than 200 lb would be required to lift a 200 lb load.

This relationship may be expressed as a ratio, as follows:

$$\text{Weight of load}/\text{Force to suspend load}$$

$$200\,\text{lb}/200\,\text{lb}$$

$$1/1$$

Therefore, the ratio is 1:1

To move the load, a force greater than 200 lb would need to be applied.

On a practical level, the direct-pull method is seldom a viable approach. Practically speaking, it would be very difficult for a 200 lb technician to lift a 200 lb load simply by pulling on the end of the rope. Instead, pulleys and ropes may be used to create a haul system with a specified MA, thereby reducing the amount of effort that a technician must exert to move a load.

The amount of MA produced by a given arrangement is calculated by formulas and is expressed as a ratio. For example, a system wherein a 200 lb load could be moved with a calculated 100 lb of input effort would be known as a 2:1 (two-to-one) MA. This calculated MA is referred to as Theoretical Mechanical Advantage (TMA) because it ignores any opposing friction in the system.

Of course, the Actual Mechanical Advantage (AMA) of the system would be somewhat less than this calculated TMA due to inefficiencies of friction, angles, and other factors. These things can be somewhat difficult to quantify, and should be fairly minor in well-constructed haul systems, so for discussion purposes in this book we will discuss hauling systems in terms of TMA.

The simplest means of reducing the amount of effort that must be exerted to move a load is to simply introduce a single pulley at the load itself. A pulley in this position also reduces the amount of force applied at the anchor (Figure 6-15).

With one end of the rope anchored, the other end is passed through a pulley at the load and the technician pulls back toward the anchor. A pulley in this position is known as a "traveling pulley" because it travels with the load. To determine the TMA in a system, simply count the number of lines that are supporting the load. With the 200 lb load supported by two strands of rope, one attached to the anchor and the other hauled by the technican(s), the person hauling feels only half the load, or 100 lb.

Expressing the load information in this second example as a ratio looks like this:

$$\text{Weight of load}/\text{Force it takes to move load}$$

$$200\,\text{lb}/100\,\text{lb}$$

$$2/1$$

Therefore, the ratio is 2:1

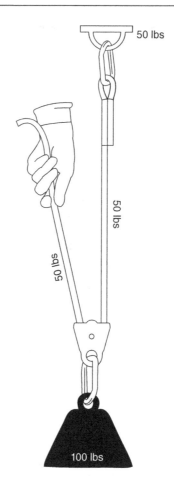

FIGURE 6-15
2:1 mechanical advantage. Illustration by Margaret Deluca.

This ratio, then, becomes the means by which we express the MA in the system. It is said to be a 2:1 MA.

It is notable that although the amount of *effort* required to move the load is *decreased*, the length of *rope* the technician has to pull through the system is *increased*. Using this 2:1 MA system, the technician must pull 20 ft of rope through the system to move the load by just 10 ft.

A second pulley added to this system at the anchor to re-direct the rope, as shown in Figure 6-16, will not offer additional MA, but it may make the direction of pull more convenient for the rescuer(s).

> To quickly determine the amount of mechanical advantage in a simple mechanical advantage system, count the number of lines actually supporting the load. In Figure 6-16 only two of the lines actually support the load; therefore, it is a 2:1 mechanical advantage. The third line, because it provides only a change of direction, is not counted as denoting mechanical advantage.

When the pulley is anchored in a stationary or unmoving position we call it a change of direction, or a fixed pulley. A fixed pulley does not offer MA. Fixed pulleys may make work easier by changing the direction of the applied force, but the same force is needed to move the load. Use caution in these situations to not overload equipment and anchorages as a result of force multiplication, discussed earlier in this chapter.

USING THE PRINCIPLES

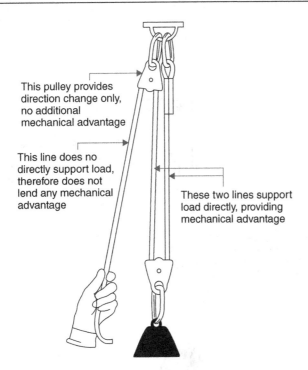

FIGURE 6-16
2:1 with change of direction. Illustration by Margaret DeLuca.

Carrying the concept one step further, if yet another pulley is added at the load end of this system, that pulley would become a traveling pulley and would increase MA. In this case, you can count four lines supporting the load, therefore the TMA would be 4:1, as illustrated in Figure 6-17. Moving our 200 lb load will require only 200/4, or 50 lb, of pull. A fourth pulley added to this system would once again provide only a change of direction, a fifth pulley would create additional TMA (6:1), and so on.

Of course, at some point it becomes inefficient to continue adding pulleys in this manner, because of the amount of internal friction within the pulleys, and also because the distance that the load moves with each reset is reduced by every pulley. For a 10:1 simple MA, one would need to haul 10 ft of rope to move the load by just 1 ft! In truth, we seldom see more than a 4:1 MA in this simple, even-numbered MA configuration.

> Whenever the first pulley in a simple system is placed at the load, the result is an even-numbered mechanical advantage. When the first pulley in a haul system is placed at the anchor, the result is an odd-numbered mechanical advantage.

Up to this point, you may have noticed that all of the MA systems we have discussed are even-numbered ratios (2:1, 4:1, 6:1). Odd-numbered MAs are also possible, as shown in Figure 6-18.

To create an odd-numbered MA, the end of the rope is tied off at the load, and the first pulley in the system is placed at the anchor. This first pulley serves only as a direction change, while the MA is introduced by the traveling pulley – in this case, the second pulley in the system.

By counting the lines that directly support the load, this can be quantified as a 3:1 MA, sometimes called a Z-Rig. For a given amount of force exerted on the system, you will realize three times that force at the load. So, to move our 200 lb load requires

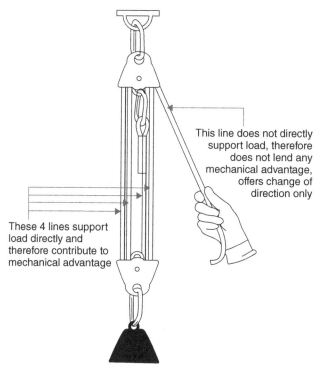

FIGURE 6-17
4:1 mechanical advantage. Illustration by Margaret DeLuca.

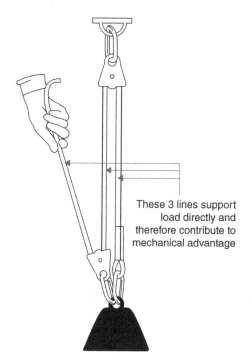

FIGURE 6-18
An odd-numbered 3:1 mechanical advantage. Illustration by Margaret DeLuca.

USING THE PRINCIPLES

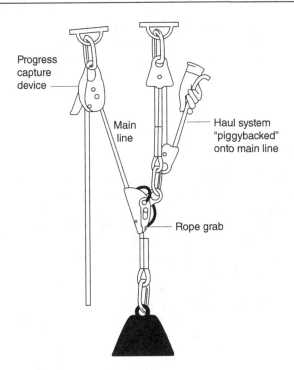

FIGURE 6-19
Piggy-back haul system with PCD. Illustration by Margaret DeLuca.

just over 66 lb of force, and for every foot that the load moves, the rescuer(s) must pull 3 ft of rope.

All of the above-described systems are "simple mechanical advantage." Stacking multiple systems on top of one another or integrating multiple systems into one another creates compound or complex MA systems. Such systems are sometimes used, but are beyond the scope of this book.

It is also notable that in most cases a haul system will not be built directly into the main working line. It is usually much more efficient to build the haul system with separate components, and then attach it to the main working line with a rope grab as shown in Figure 6-19.

Creating MA is just one aspect of a raising system. As you pull the load upward, the pulleys will move toward one another until eventually they are bunched together. Unless the load is to be raised only a very short distance, several re-sets will most likely be required. To accommodate resets, you will need to devise some method of capturing the progress gained during each haul evolution. There are several options for integrating a progress capture device (PCD) into the system.

One method to achieve this in a piggy-backed system, such as that shown in Figure 6-19, is to place a rope grab on the main line at the anchor and re-set it with each pull. This is known as PCD, or simply as a "ratchet." This progress-capture rope grab can also serve double-duty as a backup-safety mechanism in the event that the person(s) hauling should let go.

Load Ratios

Every rope access system should be built to withstand forces greater than the potential forces to which it will be subjected. Understanding the difference between system strengths and maximum potential load provides an essential foundation for evaluating system safety.

Calculation of system load ratios requires an understanding of the strength of each component, awareness of how rigging and other factors influence strength, the effect of geometry in the system, and the nature by which forces are exerted in the system. From this information, the technician will determine the load ratio – that is, the difference between the strength of each point in a given system as compared with the load that will be exerted at each point in that system.

This idea is most easily illustrated as it relates to a component, as shown in Figure 6-20.

Safety Factors

While the concept of a load ratio can be applied either to a component or to a system, from a practical perspective it is most relevant as it relates to the system. It is only by understanding the actual strength of each component *as it is interconnected with other components* in the system that one can determine the true *safety factor* of a system. The safety factor is the ratio between the weakest point in the system and the maximum potential load that is reasonably foreseeable at that point.

Rope access involves the use of two systems simultaneously: the access system and the backup system. Each of these systems will have its own safety factor, and they will most likely be quite different numbers.

Some technicians erroneously believe that if all the components in their system are rated to a strength 10× greater than the load they expect to place on the system, their safety factor of the system will be 10:1. However, this is simply not true.

A system is only as strong as its weakest link, but the weakest link in a system may not be – in fact probably isn't – an individual component at all. Rated equipment strengths are affected by rigging methods, quality of connections, and force multipliers both within and on the system. For example, the rated breaking strength of a 11 mm diameter rope may be over 6000 lbf, but when a knot is tied in that rope the strength may be reduced by as much as 30%, to about 4200 lbf. A similar reduction in strength can be assumed for a carabiner when it is three-way loaded. Often the weakest point in a system will be the interface between two components, or between a component and an external force (such as an edge).

On the other side of the equation, as we have noted, forces imparted by a load may be increased through friction, angles, and even the mechanics of the systems themselves. When a system is at rest, or unmoving, the forces remain constant. Rope access technicians call this a "static load." A load in motion (a "dynamic load") results in kinetic energy, which results in changing forces. Kinetic energy is (average) force times distance. For example, the weight of a technician with equipment may be only about 225 lb, but actions such as a free fall, or even the work that the technician is doing, can influence kinetic energy and result in potential forces up to two times the weight of the load, even without any catastrophic incident.

The dynamic energy in a falling object at the impact moment can be expressed as shown in Figure 6-21.

Considering all of this, the Competent Person technician must be capable of estimating the maximum anticipated load at any given point in the system and comparing that with the strength of the system at that point.

Given a load of 300# suspended by an anchor strap, rope, and carabiner rated at 6000#, 6000#, and 5000# respectively, we can compute load ratios as shown in Figure 6-22:

Rope strength	:	Weight of the load
6000 lbf	:	600 lbs
10	:	1

FIGURE 6-20
Component load ratio.

USING THE PRINCIPLES

$$E = F_w h$$
$$= m g h \quad (4)$$

where

F_w = Force due to gravity (weight)

g = Acceleration of gravity (9.81 m/s², 32.17405 ft/s²)

h = Falling height (m)

FIGURE 6-21
Dynamic energy.

Strap	6000/300	=	20:1
Rope	6000/300	=	20:1
Connector	5000/300	=	16:1

FIGURE 6-22
Load ratios illustrated.

However, as we have already noted, we can generally predict that the weak points in the system are the connections between the components. With this in mind, if we assume an approximate 30% reduction in the strength of the rope as a result of the knot, and about the same to the carabiner as a result of the directional loading on the carabiner, our calculations would look more like the figures shown in Figure 6-23:

Strap	6000/300		=		20:1
Rope	6000-30%(1800)	=	4200/300	=	14:1
Connector	5000-30%(1500)	=	3500/300	=	11:1

FIGURE 6-23
Load ratio assumptions corrected for rigging.

This calculation may be reasonable for the primary suspension system, or the access system, but it is probably not reasonable for the backup system. Because dynamic forces can be significantly higher than static forces, most occupational health and safety professionals recommend striving toward backup systems that will mitigate potential forces to 900# in the event of a fall. Therefore, our calculations would look more like those shown in Figure 6-24:

Here it is evident that the load ratio is much lower because of the higher potential force that must be taken into consideration.

Load ratios throughout the system will vary, so the technician should consider the ratio between the breaking strength of each point of a system and the maximum anticipated load at each corresponding point. The position at which the load ratio is narrowest dictates the safety factor for the entire system.

A system safety factor may be calculated using the following formula:

$$S/L$$

where S = Strength of the weakest point in the system and L = Maximum anticipated load

---

There is no magic formula that dictates what the safety factor should be for a rope access system. Every system should be built to withstand greater potential force than the actual force expected on the system. How much stronger will depend on the situation.

FIGURE 6-24
Load ratio assumptions corrected for rigging and maximum impact force.

Strap	6000/900	=	6.6:1
Rope	4200/900	=	4.6:1
Connector	3500/900	=	3.8:1

The target safety factor for a given operation should be established by a qualified or competent person, based on circumstances. This is one case where a higher number is not always necessarily better. The goal is to achieve as high a system safety factor as is appropriate and reasonable given the time, equipment, site conditions, and hazards. Mandating an extremely high safety factor could backfire by making equipment unnecessarily heavy, unwieldy, or incompatible with other equipment or techniques, which could actually result in increasing the overall hazard and potentially increase the probability of failure.

Probability of failure and consequence of failure are important considerations in determining an appropriate safety factor. A situation where failure is unlikely and the consequence of a failure, even if it did occur, would be minimal, may justify a very low safety factor. In contrast, a system that is subject to known conditions that are more likely to cause it to fail and/or where a system failure is likely to be catastrophic to the personnel involved, a higher safety factor may be in order.

The target safety factor should be determined first, and the system built to achieve that target. If necessary, the safety factor of a given system can be easily altered either by adjusting the system itself to use stronger components or improve the relationships between components, or by resolving to minimize loading potential on the system (i.e., one person instead of two, shorter fall potentials, etc.)

> Probability and consequence of failure should be considered in determining the target safety factor. The target safety factor should be determined first, and the system built to achieve that target.

Safety Factors are not an exact science, and should be determined by competent personnel who are specifically trained in rope access and who are capable of accurately estimating the maximum load, pinpointing the weakest link, and determining potential consequences should failure occur.

6-3 SUMMARY

A rope access technician must have a firm grasp on the principles of force, gravity, friction, and angles, and how all of these play into the systems that are created. These influences are present and are active in virtually every rope access system, and safety depends on the ability of the technicians to perform their task safely and effectively with these considerations in mind.

CHAPTER 7

Rope Terminations and Anchorages

By the end of this chapter, you should expect to understand:

- The parts of a rope
- Manufactured terminations
- Knot terminology
- The purpose of a stopper knot
- How to set and finish a knot
- Termination (of rope)
- Knot efficiencies
- Back tie
- Anchorage terminology
- The methods of connecting to an anchorage
- The purpose and methods of load sharing anchors
- Reanchor
- Deviation.

In every rope access system, rope is an integral part of the work undertaken by the rope access technician. The usability of a rope is greatly enhanced through efficient and effective termination. Terminations facilitate connection of the rope in a variety of circumstances for load management, redirection, and anchoring, among other things.

All rope systems must incorporate a solid, well-constructed anchorage as a fundamental element. The anchorage is typically a fixed structural member that offers sufficient stability and strength to serve as the termination point of the system, and to hold the system in place. Connection to the anchorage requires some manner of termination.

To effectively discuss rope terminations and anchor systems, it is essential to first establish a good baseline of accurate terms. The terminologies used in this text reflect the commonly accepted vocabulary used in regulatory language related to rope access and fall protection in the United States. Other terminologies may be

---

Professional Rope Access: A Guide To Working Safely at Height, First Edition. Loui McCurley.
© 2016 John Wiley & Sons, Inc. Published 2016 by John Wiley & Sons, Inc.

accepted, and/or definitions may vary in other jurisdictions, but the following terms will at least provide a foundation for discussion throughout this text.

7-1 ROPE AND KNOT TERMINOLOGY

The *running end* of the rope is the part of the rope used to perform work. This is also often called the *working end* of the rope, illustrated in Figure 7-1.

The *standing part* (or *end*) of the rope is the section not actively used to form the knot or rigging, as illustrated in Figure 7-1.

A *line* is a rope in use. For example, a rope used to rappel is called a *rappel line*.

A *bight* of rope is formed when the rope takes a U-turn on itself so that the running end and standing end run parallel to each another. The U portion, where the rope bends, is referred to as the *bight*, illustrated in Figure 7-2a. Many knots that form a loop from a bight in the standing part of the rope are referred as *something on a bight*, (e.g., Figure-8 on a bight).

A *loop* of rope is made by crossing a portion of the standing end over or under the running end, illustrated in Figure 7-2b.

The *tail* of a rope is usually the short unused length of rope that is leftover when a knot is tied. *Turn* refers to wrapping the rope around something and a *tuck* is to insert a part of the rope into a loop.

Anchorage Related Terminology (Figure 7-3)

Anchor Point (see also Anchorage) A place, fixing or fixture that supports and to which the various ropes and rope systems are attached.

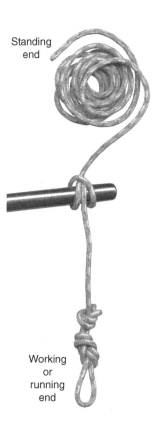

FIGURE 7-1
Parts of a rope in use.

ROPE AND KNOT TERMINOLOGY   117

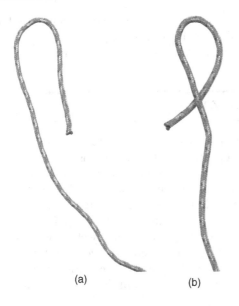

FIGURE 7-2
(a) bight (b) loop.

FIGURE 7-3
Anchorage-related terminology.

Anchorage (see also Anchor Point) A place or fixture that supports and to which the various ropes and rope systems are attached.

Anchorage Connector The means of securing to an anchorage, such as a strap, eye bolt, and so on.

Anchor System The combined anchorage and anchorage connector, working together to support a load.

It is the term "Anchor," when used alone as a noun, that can be most misleading, and the confusion is usually whether the term might mean a mere anchor point or a full-fledged anchor system. To avoid this kind of confusion, it is best to simply avoid using the word "anchor" as a stand-alone noun. The term "anchor" may be appropriately used as a verb in common rigging vernacular, that is, "*to anchor the rope,*" and indeed you may find it used in this manner in this book.

7-2 ROPE TERMINATIONS

The rope may be terminated by the manufacturer prior to sale, or by the technician in the field. Manufacturer-provided terminations include pretied knots, sewn eyes, and swaged ends, but such terminations are by definition of limited value. Field terminations are more versatile and useable, as their nature and location may be customized to the need at hand. These most often consist of knots, although certain hardware components are also available to facilitate field terminations of rope.

While some industries may philosophically object to the use of knots, they are an essential entity to rope access. The elemental versatility of rope access is facilitated through the use of knots, which in turn increases the safety of the technician. For this reason, it is important that technicians understand the proper use and limitations of each.

7-3 MANUFACTURED TERMINATIONS

Most terminations will decrease the strength of a rope to at least some degree. One notable exception is the splice, used in braided and laid rope, which in some cases is actually stronger than the strength of the rope itself. However, splices do not provide practical means to terminate modern kernmantle life safety rope. Swages and sewn terminations, shown in Figures 7-4 and 7-5, respectively, are the most common types of manufacturer-provided terminations in life safety rope, although some manufacturers will also provide pretied knots. When a manufacturer provides a terminated rope, it should be appropriately marked with the terminated strength. If technicians note that the rated strength of a terminated rope is the same as the rated strength of that same rope without terminations, they should ask for clarification and testing results from the manufacturer.

FIGURE 7-4
Swaged termination.

KNOTS

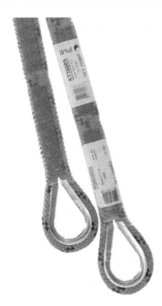

FIGURE 7-5
Sewn termination.

One disadvantage of manufacturer-provided terminations is that the termination remains in the same position throughout the life of the rope. When a material bends repeatedly in the same location, over time it can become weaker. Known as flex-fatigue, this condition can be offset or avoided simply by changing the location of the termination, even slightly, with each use.

Permanent terminations should be inspected carefully prior to use, and the rope should be retired if the rope becomes particularly soft or floppy at the point where the rope is terminated.

7-4 KNOTS

Knots and ties, whether tied by the manufacturer or the technician, will also reduce the strength of the rope at the point where it is tied. In this text we will discuss six basic types of ties, based on their functions: (i) stopper knots, (ii) end-of-line knots, (iii) midline knots, (iv) knots that join two ropes, (v) safety knots, and (vi) hitches.

When working with ropes, it is critical to be aware of the type of material into which the knot is tied. Some fibers have a low coefficient of friction and require special considerations when tying knots, and some fibers have been shown to result in different knot efficiencies. Knots that are effective on a rope do not always perform well in webbing or sling material.

Stopper Knots

A stopper knot is a type of knot tied in the running end of a rope to prevent the practitioner from descending off the end. Stopper knots may also be used elsewhere in the system when a stopper function is warranted.

The *overhand knot* is a simple tie that may be used for this purpose, but its relatively low bulk can limit its effectiveness for this purpose. To tie an overhand, simply make a loop in the rope, tuck an end of the rope through the loop from behind, and pull taut, as shown in Figure 7-6

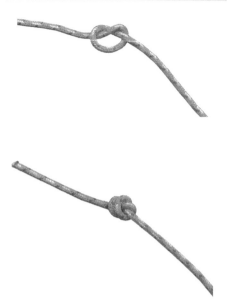

FIGURE 7-6
Overhand knot.

FIGURE 7-7
Barrel knot.

With a couple of more wraps, the overhand becomes a *barrel knot* (Figure 7-7) – somewhat bulkier and thus better suited as a stopper. To tie the barrel knot, begin as though beginning the overhand, but then take the end of the rope through the loop a couple more times before pulling it tight. This knot works well as a stopper because it is large, and does not roll or slip when tightened correctly.

Another common stopper knot is a *figure 8 stopper knot*, shown in Figure 7-8.

This figure 8 also forms the basis for other useful knots including the figure-8 Retrace and figure 8 on a bight. To tie a figure-8 stopper knot, the running end of the rope is passed completely around the standing part as though for an overhand, but is made to continue an additional half-turn around the rope before putting the running end through the loop from the front, and tension.

When two ropes are used in tandem, the stopper knot may be tied into each line individually, or into both lines together, as shown in Figure 7-9.

Stopper knots – basic although they may seem – provide the foundation for many of the other knots used in rope access so it is important to learn these before progressing further.

The following discussion provides an overview of several different types of knots that are considered appropriate for rope access. However, step-by-step instructions are not included in this text.

End-of-Line Knots

In its simplest form, a *loop* is a section of rope that crosses itself. The term *tied loop* refers to a knot that forms a fixed eye, or loop. Such a loop may be tied into the end of a rope to anchor, tie in, or attach the rope to something.

The bowline, a knot borrowed from mariners, is familiar to many; however, this knot can "capsize" into a slipknot quite easily when the tail is pulled. For this reason, the *high-strength bowline knot* (Figure 7-10) is considered by many to be a more appropriate choice for life safety applications.

Other variations of the bowline knot that include added safety for live loads are the simple *bowline with safety* and the *bowline with Yosemite safety*. These are not addressed in this book. Another interesting bowline variation is the *triple loop bowline*

FIGURE 7-8
Figure-8 stopper knot.

(Figure 7-11). This knot is tied in the same way as a basic bowline, but on a bight of rope rather than on a single strand. The tail is then extended to create a third loop. This knot can be tied in the middle of a line, and its three adjustable large loops can be used creatively – for example, to create a load-distributing anchor.

Perhaps the most common knots used to form loops in rope access are the various renditions of the figure 8. You've already learned the basis for this knot as a stopper, but the figure 8 is also a good choice to form a bight (Figure 7-12) or even a double-bight (Figure 7-13) in the end of a rope. Other variations of this knot (shown later in this chapter) may be used midline and even to join two ropes.

Although the versatile nature of the figure-8 knot is indeed attractive, it should be noted that learning only one knot may prove limiting. Some people feel that the figure-8 knot is easier to tie and check than other knots. This knot involves a second strand of rope following the first strand along a parallel path, but some feel that the redundant nature of the figure-8 retrace can make it deceptive on visual inspection. When one strand of rope appears to be correctly tied, the brain can recognize it as a figure-8 knot and may not emphatically register that the second strand is missing. This has been reported as a contributing factor in some accidents (Gonzales, 2008).

Perhaps the easiest and most common method of throwing a loop into the end of a rope is with a simple *overhand on a bight* (Figure 7-14). However, this knot is seldom used as it does not offer much redundancy, and can be quite difficult to untie after use.

FIGURE 7-9
Stopper knot in 2 lines.

FIGURE 7-10
High strength bowline.

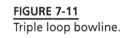

FIGURE 7-11
Triple loop bowline.

FIGURE 7-12
Figure 8 – single loop.

FIGURE 7-13
Figure 8 – double loop.

FIGURE 7-14
Overhand loop.

Midline Knots

It is often desirable to form a loop in the middle of a rope for clipping into or grasping it, or for bypassing a piece of damaged rope. While the end knots described above may be possible to tie in the middle of a rope, they can be susceptible to rolling or deforming when not loaded properly. When a loop is created at the end of a rope, it may be assumed that the standing part of the rope will be anchored, and that a load may be applied to the midline loop itself or to the running end below the loop. With most end-knots, if the rope below the knot is loaded, the knot deforms and weakens.

FIGURE 7-15
Inline figure-8.

For this reason, knots that are intended to be loaded in any of the potential directions should be selected.

The *inline figure-8* (Figure 7-15) is one example of a midline knot that provides a straighter-line pull and is much more appropriate for this purpose.

Perhaps the most favored midline solution is the *butterfly knot* (Figure 7-16). This knot is a particularly good choice when the loop and the line beneath it are to be placed under significant load. The butterfly knot can be pulled effectively either from the loop or from below the knot without negative effect, and it is relatively easy to untie even after loading. Caution must be taken with this knot, because if the loop is not big enough or if it is not loaded, it can pull out under tension.

Knots (Bends) That Join Two Ropes

A *bend* is a type of tie that connects two rope ends together. Tying a knot that will not untie itself is very important when joining two ropes, particularly because the knotted ends are unlikely to be in a place where they can be constantly monitored.

FIGURE 7-16
Butterfly knot.

FIGURE 7-17
Double fisherman's bend.

One secure and low profile option for this purpose is the *double fisherman's bend* (Figure 7-17) This bend is very effective for joining ropes of relatively equal diameter. Care should be taken to ensure that the two halves of the bend nestle against each other and that there is enough tail protruding from the knot to keep the knot from unraveling. This bend is also commonly used to join two ends of a short length of cordage into a loop.

When ropes of unequal diameter are joined, the *double-sheet bend* (Figure 7-18) is a more effective tie. This is a bulkier alternative that is perhaps not quite as strong, but can be easier to untie and is preferred for joining ropes of different diameters.

The versatile figure-8 knot also deserves honorable mention here. Retracing a figure-8 knot in the opposing direction with a second rope results in what is commonly referred to as a *flemish bend*, an effective means of joining rope ends

FIGURE 7-18
Double-sheet bend.

FIGURE 7-19
Flemish bend joins 2 ropes.

(Figure 7-19). Care must be taken with this method to ensure that the rope ends exit from opposite ends of the bend. If tied simply as a figure-8 knot, this bend has a tendency to deform and pull itself apart.

Flat webbing and similar materials perform differently under tension. The preferred bend for joining webbing ends is known as the *ring bend*, sometimes also called the *water knot* (Figure 7-20). This is most useful for forming webbing slings into a loop, but can also be used for ropes and lashing.

Knot Safety

A knot is not finished until it is properly dressed. Dressing a knot is the process of straightening, adjusting, and snugging the knot into its proper shape so that it will load properly. Once dressed, the knot should be "set" – that is, pretensioned with enough force to firmly settle it into place. If a knot is not adequately dressed and set after tying, it can deform when it is loaded.

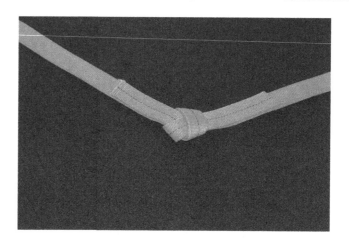

FIGURE 7-20
Ring bend (water knot). Courtesy of Robert McCurley.

FIGURE 7-21
Overhand knot as safety.

Every knot should be checked (preferably by someone other than the person who tied it) to ensure that it is tied properly, and should be monitored at intervals thereafter. Many knots have a tendency to loosen, and some can even change forms (e.g., into a slipknot).

A safety knot can help prevent mishaps. A safety knot is an overhand knot (see Figure 7-21) or *barrel knot* (Figure 7-22) tied into the tail of the rope after a knot is tied.

The safety knot should be positioned very close to the primary knot to help prevent the primary knot from deforming or unraveling.

Hitches

A knot that is tied around something (e.g., a tree, a standing rope, the rail of a litter) and conforms to the shape of the object around which it is tied, and that does not keep its shape when the object around which it is tied is removed, is actually not a knot but is more appropriately called a *hitch*. Hitching is a method of tying a rope around itself or an object in such a way that the object is integral to the support of the hitch. Hitches are used occasionally in rope access and should be considered for use only by a skilled technician, because there are severe consequences when a hitch

FIGURE 7-22
Barrel knot as safety.

comes untied or does not perform as intended. Specifically, disintegration of a hitch results in the immediate release of whatever load it is holding.

One of the most commonly used hitches is the *Prusik hitch* (Figure 7-23). A Prusik hitch is a sliding friction hitch by which a cord can be attached to a rope and slid up and down the rope for positioning. However, under tension, the hitch will not slide. A looped length of cordage intended for this purpose is known as a Prusik loop. Commercially available Prusik Loops are typically sewn into a loop, but a Prusik loop may also be created by tying a length of cordage into a loop by means of a double fisherman's bend. Wrapping the loop around the main rope and through its own loop two or three times and then pulling it tight forms the hitch. Prusik hitches are not permitted for fall protection in industry, but they can be useful to know for rigging purposes and are handy tools for emergencies.

Another fairly common hitch used in rope access is the *clove hitch* (Figure 7-24). This hitch can be useful when trying to shorten the distance between two objects. It is also useful in some lashing techniques. One nice thing about the clove hitch is that it can be adjusted even while under load but it should be used with care as it can also have a tendency to roll loose.

Another type of hitch, called the *Münter hitch* or Italian hitch (Figure 7-25), can be used around a carabiner or pole to add friction to a system, as in a belay. This hitch is particularly useful in that it effectively adds friction regardless of which direction the rope is moving. However, care should be taken when using the hitch around a carabiner, because there can be a tendency for the moving rope to slip through the gate of the carabiner, rendering the hitch useless.

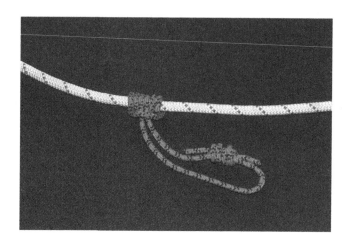

FIGURE 7-23
Prusik hitch. Courtesy of Robert McCurley.

FIGURE 7-24
Clove hitch.

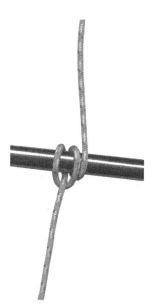

FIGURE 7-25
Munter hitch.

KNOTS 131

FIGURE 7-26
Truckers hitch.

FIGURE 7-27
Girth hitch.

A very handy hitch known as a *trucker's hitch* (Figure 7-26) is useful for pulling cord or webbing tight across something (e.g., a load in the bed of a pickup truck [hence the name]) or tensioning a back-tied anchor.

No discussion of knots would be complete without mention of the *girth hitch* (Figure 7-27). Also known as the Lark's foot, this hitch is useful for a quick, although not too secure, attachment of a sling or rope to almost anything.

FIGURE 7-28
Knot efficiencies.

Knot	Retained efficiency
Double fisherman's bend	65–70%
Bowline (single)	70–75%
Figure-8 on a bight	75%–80%

7-5 KNOTS AND ROPE STRENGTH

The strength of a knot is directly proportional to the strength of the material in which it is tied. Knot strengths are usually expressed in terms of *efficiency ratio*. A knot rated at 85% efficiency is said to maintain about 85% of the reported breaking strength of the rope.

Some individuals and agencies have reported that any knot reduces the strength of a rope by at least 50%. This information is erroneous, because the efficiency of any knot depends on which knot is used, which rope it is tied in, whether it is tied correctly, and how it is maintained. Most known and accepted knots reduce the strength of a typical life safety rope by no more than 30%. That is to say, the retained strength of the rope even with a knot in it is generally no less than 70% of that rope's minimum breaking strength. The most commonly used knots are even more efficient and in the range of 80%.

Unfortunately, accurate data about knot efficiencies are hard to find. Comprehensive testing that takes into account statistically significant sampling, differences among rope fibers, constructions, and diameters, static versus dynamic loading, and other variables is virtually nonexistent.

The data shown in Figure 7-28, derived from knot-testing results presented at the International Technical Rescue Symposium (ITRS) and subsequently published,[1] reflect a range of estimates and should be referenced for trend information only.

The best way to know the strength of a knot on a given rope is to test the type of knot that is to be used on the type of rope that is to be used. One should test enough samples using statistical analysis to provide a reasonable margin of error.

The above referenced research, which was authorized by the Cordage Institute, shows that the use of appropriate knots in certain rope-based life safety systems, using polyester and nylon ropes, is a safe and acceptable practice (Richards, 2004). In these tests, four common knots were pull-tested to determine the efficiency. (The term "efficiency" refers to the relative strength as a percentage of unknotted break strength.) The samples were *slow-pull tested to CI 1801 test specifications, on a wide range of diameters of kernmantle life safety ropes. The results were compared with tests of unknotted ropes from the same respective batches.

7-6 ANCHORAGES

Any anchorage should be capable of withstanding the maximum potential force in any direction and to any magnitude that may be reasonably foreseeable. One

---

[1] Dave Richards (2005). "Knot Break Strength vs Rope Break Strength". *Nylon Highway* (Vertical Section of the National Speleological Society) (50). Retrieved 2010-10-11.

method of predicting the direction of pull involves following the fall line. This can be assumed to be the direction that a load would take from the point of the anchorage, given the pull of gravity and the influence of terrain and other factors.

Classifications of Anchorages

Most anchorages are classified as being either structural or installed. A structural anchorage is one that is formed by an existing structure, which is either natural or manmade. In the natural setting, structural anchorages might consist of a tree or a large boulder. In manmade environments, structural anchorages may include building elements and supports.

Generally, structural anchorages can be an excellent choice for ropework (Figure 7-29), but care must be taken to verify the security and appropriateness of the structure for such use. Often the thing used as a structural anchorage would not have been originally designed or intended for this purpose, but somewhere along the way the element would have been deemed appropriate by a person qualified to make such a determination. In some cases, special attachment points may be secured into an existing structure expressly for this purpose.

Installed anchorages, Figure 7-30, are those that are placed expressly for the purpose of being an anchor. They may be permanently installed or temporary in nature. A tripod over a hole, a removable bolt, and an eye-bolt, are all examples of installed anchorages but even from these simple examples you can see that the delineation between an anchorage and an anchorage connector is not always clear.

Philosophical discussions aside, what really matters is the performance of the anchorage system as a whole. Preplanned permanent anchorages are advisable in locations where work at height or rescue is frequent. Such an anchorage is normally referred to as an engineered anchorage, and in some cases it may even be required that a registered Professional Engineer perform the installation and/or offer certification of the anchor.

In selecting an anchorage for a given application, the technician must consider both the direction and the magnitude of potential forces that are likely to be generated by reasonably foreseeable loading. Effects from any possible deflection of the anchorage, and the impact on the structural members to which the system is attached are also important considerations.

FIGURE 7-29
Any structural element may be considered an anchorage. Courtesy of ClimbTech.

FIGURE 7-30
An installed anchorage. Courtesy of ClimbTech.

Anchorage System Performance

A rope access system comprises two subsystems: an access system and a backup system. Each of these systems should be anchored independently from the other, although it is not uncommon to make a connection from the access system to the backup anchor, and vice-versa, just for added security.

Because the access system and the backup system in rope access are fully interchangeable, the technician may choose at any given time to switch between the two to accommodate a particular maneuver or goal. For this reason, both anchor systems must be capable of withstanding the potential forces of either application.

The industry standard for anchor strength is 5000 lb, which stems from OSHA's 1910.66 requirement for personal fall arrest systems (although anchors designed by a qualified person are permitted to have a reduced strength, as long as a safety factor of at least two is maintained). Each rope, whether for fall protection or work positioning, should be independently connected to its own anchor that should be capable of holding a 5000lb load without failing. Other (nonregulatory) organizations provide similar guidance on anchor strength. The American National Standards Institute (ANSI) has, for the most part, taken the 5000 lb stance on fall arrest system anchors, and allows "certified" anchors to have breaking strengths reduced to twice the maximum permissible impact force.

According to current industry best practice, noncertified rope access anchorages should be selected to withstand at least 5000 lbf (per person attached) while certified anchorages should be rated to at least two times the maximum potential arrest force (in event of a fall) multiplied by the number of persons attached. Only a qualified person, generally a professional engineer, may establish a certified anchorage.

ANSI Z359 is the code of practice most often followed by those who work at height in the United States. This document calls for anchorage strength requirements for noncertified anchorages as outlined in Figure 7-31.

Rope access is characterized, at least in part, by the ability of the technician to rescue his partner from typical suspended rope predicaments. While nonentry methods are preferred, coworker-assisted rescue that involves direct contact may require that two technicians are on a given system at the same time. In such a case, where anchorages may have to sustain a two-person load, it is recommended that anchorages be used that are rated to a strength of at least 3000 lbf per person attached (or, 6000 lb), in accordance with the ANSI Z359 requirements.

ANCHORAGES

Intended use	Minimum strength	Multiply by
Restraint	1000# (4.5 kN)	# Authorized connections
Positioning	3000# (13.3 kN)	# Authorized connections
Rope access	5000# (22.2 kN)	# Authorized connections
Vertical lifeline	5000# (22.2 kN)	# Authorized connections
Rescue	3000# (13.3 kN)	# Authorized connections

FIGURE 7-31
Z359 strength requirements for noncertified anchorages.

The strength of an anchor system is completely dependent upon the strength and configuration of the components that connect the system to the anchorage. The connecting components should be of sufficient strength for the task to be performed, and should be attached in such a way so as to prevent unwanted movement or disengagement of the system from the anchorage. Anchorage connectors are addressed further in Chapter 4, Equipment for Rigging.

Positioning the Anchorage System

Rope access anchorages should be selected to be in line with and reasonably high above the system to be used so as to facilitate edge management and reduce potential fall distances. The effects of vector forces, as discussed in Chapter 6, should be managed in terms of both actual force and direction of pull.

While conventional fall arrest systems can permit as much as a 2 m fall (or greater) as a matter of course, rope access is much more conservative and systems are rigged so that potential falls do not exceed 1 m. Yet, it is at the edge where the most significant fall potential exists. A high anchor provides the technician with security and stability when transitioning over the edge and onto the system.

When an anchor itself is not of sufficient height, or where a high anchorage is not available as near to the edge as the technician might wish, a similar effect may be achieved by establishing the anchor far back from the edge and passing the line through a high point at or near the edge. This type of setup is known as a "high directional."

High directionals may be rigged from natural features or from artificial components, as shown in Figure 7-32. Of course, introduction of such a high point questions

FIGURE 7-32
Artificial high directionals help facilitate rigging.

FIGURE 7-33
Anchorage connector with back-tie for stability. Courtesy of Trask Bradbury.

nearly all of the rigging theory discussed in Chapter 6, and great care should be taken to ensure that the resultant direction of pull does not create a tipping action on the high directional equipment used. Special training is required for the use of this equipment.

Back-Ties

A back-tie (Figure 7-33) may be incorporated into the anchor system where extra security is desirable, or where there is concern that an anchor may shift or creep during use. A back-tie is not in itself considered to be an anchorage, and it need not necessarily be of equivalent strength as that of the primary anchorage. The purpose of a back-tie is simply to provide security and stability to an anchorage.

The intricacies of back-ties are an advanced skill, and are beyond the scope of this book. The concept is noted here only so that the reader is aware and will recognize the concept if encountered in the field.

Direct Attachment

Direct attachment of a system to an anchorage occurs when a terminated line is directly connected to the anchorage point using a carabiner or other connector, as illustrated in Figure 7-34.

This example of a direct attachment shows the use of an end knot to create a bight in the end of a rope, and the bight connected to an anchorage connector using a carabiner.

When connecting directly to a fixed anchorage, particularly when connecting a hard connector to a D-ring or O-ring type anchorage, it is essential to ensure that the size and dimensional relationship between the anchorage and the connector are compatible. Some of the geometries achievable between an anchorage connection and an anchorage can result in a magnification of the applied load on portions of the anchorage connector and/or anchorage. Incompatibility between anchorages and anchorage connectors can also contribute to a concept known as rollout, where connecting components press against one another in such a way so as to cause one or the other to become dislodged, as shown in Figure 7-35. This concept was introduced in Chapter 4, but should be noted here as it is highly applicable to anchorages.

ANCHORAGES 137

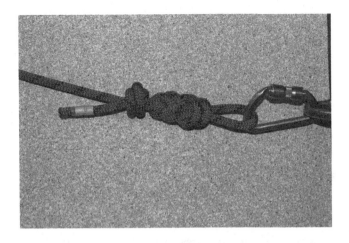

FIGURE 7-34
Direct attachment to an anchorage connector. Courtesy of Robert McCurley.

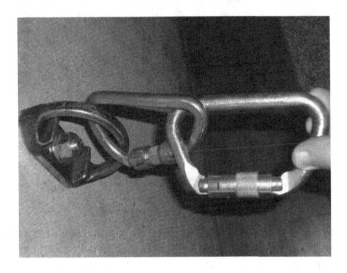

FIGURE 7-35
Avoid potential for rollout.

Another type of direct attachment occurs when a line is simply wrapped around an anchorage point and then terminated by some means – for example, by a knot. Figure 7-36 shows a line wrapped around an anchorage and tied with a knot.

The strength of the knot should be factored in when analyzing the strength of the anchorage, and the inside angles of the looped rope should be narrow enough to maintain the shape and structure of the knot and ensure good management of forces. The effects of angles as they relate to forces are discussed in further detail in Chapter 6.

Perhaps the simplest and strongest method of creating a direct connection between a fixed line and an anchorage is the tensionless anchor, illustrated in Figure 7-37.

For this, only the anchorage and a rope are required, and the security of the anchorage relies on the principle of friction. The tensionless anchor is best used on a fixed structure whose edges are not too harsh. It may be padded as needed to protect the rope or the structure. The optimum anchorage diameter is at least four times the diameter of the rope used, but not so great a diameter that it "wastes" valuable rope length.

To create a tensionless anchor, the standing end of the rope should be wrapped around the anchorage several times. The effect of gravity should be considered when

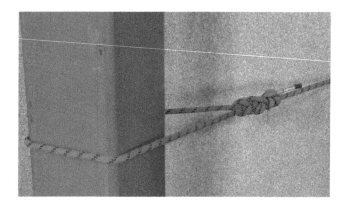

FIGURE 7-36
Line tied directly to an anchorage. Courtesy of Robert McCurley.

FIGURE 7-37
Tensionless anchor. Courtesy of Robert McCurley.

wrapping: if the rope is wrapped in an upward direction, rather than down, gravity contributes positively to the security of the connection. In practice, three to four wraps normally result in enough friction that pulling on the working end of the rope will not result in movement, but of course the diameter of the anchorage and the coefficient of friction will also have some effect. For an added measure of security, the standing end of the rope should be knotted, a carabiner clipped into it, and then the carabiner should be clipped loosely over the working line as it exits the tensionless anchor. Care should be ensured when clipping off the standing end of a tensionless anchor to avoid creating a change of direction in the working line.

A direct attachment is appropriate when the anchorage is in line with the direction of travel (fall line) and is of sufficient strength and stature to accommodate the system.

Load Sharing Anchor Systems

At times, multiple anchorage points may be combined to create an appropriate anchorage system. This occurs when the position or arrangement of a single anchorage point is not sufficient to direct or resist the desired force as desired. In this case, a load sharing anchorage, also known as a load distributing anchorage, may be used to distribute the load between two or more anchorage points. Loads are best distributed between anchorage points when the anchorages are positioned somewhat adjacent to one another and not too far apart. In the simplest form of a load sharing anchor, a single sling is placed in two anchor points, and a knot is used

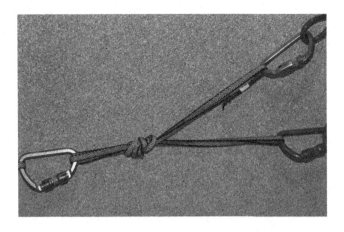

FIGURE 7-38
A fixed load sharing anchor, one sling. Courtesy of Robert McCurley.

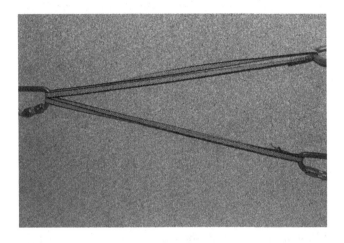

FIGURE 7-39
Fixed load sharing anchor, two slings. Courtesy of Robert McCurley.

to form a loop between the two anchors, fixing the position of the load at the knot, as shown in Figure 7-38. Using this method, the location of the knotted connecting point can be adjusted to accommodate a variety of fall-lines.

A similar result can be achieved by using two separate slings, one to each anchor point, and bringing them together to clip into the load line as shown in Figure 7-39

When the integrity of an anchorage relies on the interdependence between two or more anchorage points, as in a load sharing situation, the combined whole is typically considered to be but one anchorage. One load sharing anchorage system equals one anchorage for one line.

One potential disadvantage of a fixed load sharing anchor is that as the direction of pull changes (e.g., if the fall line varies or if the path of the technician deviates) it is possible – even likely – that the entire load might become supported by just one of the anchor points in the system. This somewhat defeats the whole concept of having created a load sharing anchorage in the first place, and should be avoided if possible.

More equitable distribution of the load in a load sharing anchor may be achieved by rigging the anchor to "self adjust." A self-adjusting load sharing anchor system can be created as illustrated in Figure 7-40 by simply connecting a sling between the two anchor points, making a twist, and clipping into the X, as shown.

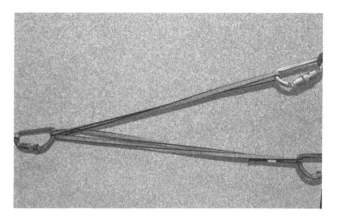

FIGURE 7-40
Self-adjusting load sharing anchor. Courtesy of Robert McCurley.

While building a robust load sharing anchor system, consideration must be given to what might happen in the unlikely event that one of the anchorage points might fail. The purpose of the twist introduced in the description above is to ensure that if one anchor point should fail, the remaining leg is "captured". Simply clipping into the bight formed between the two legs, without the twist, might result in catastrophic failure in the event that one anchor point were to fail. Potential impact forces in the event of the failure of an anchor point should also be considered.

Minimizing the length of the adjustment component in the self-adjusting load sharing anchor as shown in Figure 7-41 will help to reduce the distance and force that may be generated if one anchor point fails. This can be accomplished by extending each anchor point toward a smaller focal area. A sling made of a slippery material such as ultra high molecular weight polyethylene (UHMWP) works well to provide the adjusting function. Although UHMWP offers poor force absorption capability, the slippery nature of these fibers allow for better equalization of the load, and the shorter sling helps to prevent a fall of significant distance.

Load sharing anchors, whether fixed or self-adjusting, are not appropriate for every situation; they are but one tool available to the rope access technician. Like any option, the potential benefits of such anchorages should be considered on an as-needed basis, with consideration toward the potential disadvantages as well as the potential advantages. There is no one perfect solution for all situations; rigging requires constant vigilance, and the "best" choice for a given situation is made based on consideration of all factors.

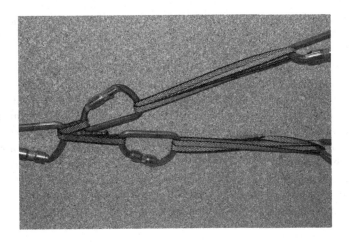

FIGURE 7-41
Extend anchorage points to reduce potential fall distance. Courtesy of Robert McCurley.

Angles in Anchor Systems

Regardless of whether anchor points are used together to form a fixed or self-adjusting version of a load sharing anchorage system, the technician must maintain awareness of the angle formed within the anchorage. Minimizing the internal angles, as discussed in Chapter 6, will help reduce the potential forces on the respective anchor points.

Angles are discussed in detail in Chapter 6, but it bears repeating here that the technician must understand how angles affect forces as they are applied to a system. In a two-point anchor system with 120° of angle between the legs, each leg sees 100% of the load, as shown in Figure 7-42.

As the angle exceeds 120°, magnification of forces on the two anchors becomes even more extreme. The angle of 120° is often referred to as the "critical angle" because beyond this range the forces can become untenable. By the time the angle

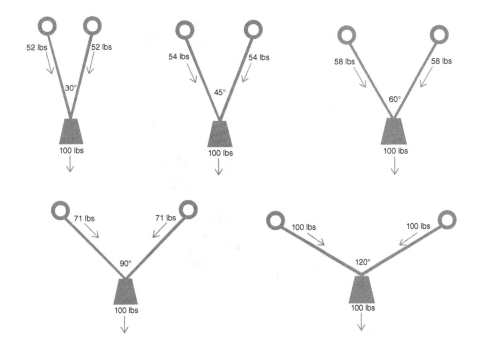

FIGURE 7-42
As angle increases, so does force on each leg of a two point anchor. Courtesy of Robert McCurley.

is at 175°, each anchor can be exposed up to 11 times the force of the load! For this reason, some technicians try to keep angles to less than 90° as a rule of thumb and 120° as an absolute maximum.

The hazard that is created in establishing a rule such as this is that responders may come to rely on the rule in lieu of really understanding the effects of the angles on the system. It is most important that responders understand how forces may be amplified, and that the anchorages and other parts of the system are able to withstand whatever forces the technician might impose (plus an appropriate factor of safety). In other words, even 120° may be too great an angle for your system, if your anchorages and components cannot all withstand 100% of the load (plus safety factor). On the other hand, it may be perfectly fine to exceed a 120° angle in an anchor system as long as your anchorages and components can all withstand several times the resulting applied force (plus safety factor).

Change of Direction

When the system or load might need to deviate from the fall line or the anticipated direction of pull, it may be appropriate to establish an additional anchor to create a change of direction.

There are two different approaches to changing direction. The first, called a **directional deviation** (Figure 7-43) is simply a midline anchor into which a line is clipped but not totally tied off. A good rule of thumb when rigging a deviation is to limit the rope angle less than 20 degrees from vertical between the deviation and main anchor. This will help mitigate potential swing fall. Another way of mitigating swing fall is to keep the horizontal offset between anchors to less than 6 ft. When distances greater than 6 ft must be negotiated, a series of deviations may be used as shown in Figure 7-43.[2]

FIGURE 7-43
Directional deviation. Courtesy of Trask Bradbury.

---

[2] *Safe Practices for Rope Access Work*. SPRAT, 2012.

SUMMARY

FIGURE 7-44
Rebelay or Reanchor. Courtesy of Pigeon Mountain Industries, Inc.

In a directional deviation, the line is allowed to run freely through a connector or pulley device. The performance requirements of an anchor system for a directional deviation will depend on the intended load, and the probability/consequence of failure. Vector force amplification should be considered, as this may necessitate strengths even higher than the original anchorage requirements. The force at a directional anchor can reach 1.4 times the force of the load with a 90° bend in the line.

Care must also be taken when using this type of a system to avoid potentially dangerous pendulum falls. The use of directional deviations is discussed in greater detail in Chapter 11.

Another approach to creating a change of direction is to re-anchor a rope midline, so that the length of the rope above is essentially independent from the rope below. This method is called a rebelay, or re-anchor (Figure 7-44). When re-anchoring a rope, sufficient slack should be left in the top length of the rope to ensure that the upper section of the rope will not become trapped or encumbered by the re-anchor during use.

A re-anchor system should be at least as strong as the original anchor, or five times the potential load – whichever is greater. Considering previous discussions regarding the effect of angles on a system, calculating the potential load on a directional deviation can be particularly challenging. Chapter 9 addresses re-anchors in greater detail in context of the full rope access system.

7-7 SUMMARY

Anchorages for use in rope access should be selected, identified, and marked as part of the preplanning process. In addition, as part of this process, the possible need for additional anchors for rescue situations should also be considered. Predicting emergencies is not an easy task, and predicting where exactly a subject might need to be lowered or raised can be even more difficult. Nevertheless, this is an important part of the work plan.

Before any anchor system is subjected to a live load, a safety check should be made to ensure that

- the anchor system is able to withstand the load;
- the overall strength of the anchorage is sufficient and offers an adequate margin of safety;
- the anchorage is secure, and that it is not likely to slip or move when loaded;
- the anchorage and anchorage connector(s) are in good condition.

All potential anchors must be thoroughly evaluated by a competent person before use to ensure that overall system safety requirements can be met. Evaluation of anchor systems is best done by experienced personnel, as this is a somewhat subjective process and requires training and experience.

CHAPTER 8

Rope Access Systems

By the end of this chapter, you should expect to understand:

- Access Zones and Hazard Zones
- Equipment Compatibility
- Vertical Systems Rigging
- Pull-Through Systems
- Edge Protection
- The Access System
- The Backup System
  - Clearance
  - Swing Fall
- Directional Deviations
- Rebelay Systems
- Welfare of the Technician.

This chapter will summarize the contents of the previous chapters into a set of concise general guidelines for rigging basic rope access systems. It builds on all of the previous chapters, and in particular on a good understanding of equipment (Chapters 4 and 5) and on rigging concepts (Chapter 6) are an essential prerequisite.

In a recent survey of several hundred SPRAT certified rope access technicians, rope access methods were reported as being used in more than 30 different kinds of jobs. Some of the most common environments where rope access was reported as being used are shown in Figure 8-1.

Many certified technicians find themselves in high demand, and use rope access in a variety of different job applications. This was reflected in the survey in that although 68% of respondents reported using rope access in conjunction with buildings, every respondent ticked off more than one selection. Applications other than buildings were selected in over 82% of all responses, suggesting that rope access is an excellent choice for a wide range of applications.

Professional Rope Access: A Guide To Working Safely at Height, First Edition. Loui McCurley.
© 2016 John Wiley & Sons, Inc. Published 2016 by John Wiley & Sons, Inc.

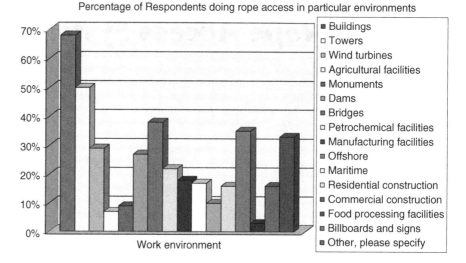

FIGURE 8-1
Common applications for rope access.

In the survey, each of the 15 different workplace-type options offered for selection was selected by one or more respondents. In addition, respondents were given the option of selecting the category of "other," and adding unlisted categories. Some of the categories written in to the "other" category included (but were not limited to) radar installations, mines, power generation facilities, entertainment rigging, theme park rides/attractions, geologic features, arboriculture, masts, antennae, water towers, avalanche control work, power lines and utilities, museums, smoke stacks, pressure pipes, highway maintenance, raptor nest removal, government facilities, oil and gas exploration, cranes, refinery plants, ski gondolas, radio telescopes, challenge course construction, wharfs, quarries, and solar panel installation.

The variety of applications in which rope access is used attests the fact that this method of safety provides key tools in the form of innovative approaches to solving challenges associated with working at height.

Rope access as employed in industry today involves far more than descent of buildings, and is in fact the access method of choice for numerous inspections and maintenance of critical infrastructure and assembly. Carefully planned and executed rope access is an appropriate application of equipment and techniques based on best practices taken from a wide range of applications (including rescue) that have been vetted and proven for many years. Rope access is recognized, effectively used, and sufficiently regulated or standardized all over the world.

If the ultimate goal is safety, the option of rope access should be considered as a viable approach to accessing work at height.

Rope access is primarily concerned with moving up and down, and working while being suspended from ropes. An effective work at height system requires suitable equipment, worker competence specific to rope access, and appropriate management. Inadequacy of any of these factors can result in catastrophic failure. Having explored the general concepts of what rope access is, and the components of equipment that are used for rope access, the next step is to delve into worker competence. Before we discuss the specific techniques and methods required for performing the requisite skills, it is important to have a firm grasp on how the foundational systems are rigged for these purposes.

Any rope access technician who is appropriately trained to do so may prepare and install rigging. A thorough understanding of rigging is an important part of any technician's training, even if it is at the most basic level. Systems should always be checked and verified for safety before use by a competent person other than the person who performed the rigging, and in addition to this it is in the best interest of any technician who is about to get on rope to inspect the system they are about to engage in. Inspecting a system for safety should include more than just a cursory visual check. Equipment should be specifically verified to be rigged properly for function, and sufficient tension should be applied to the system to ensure that it will load properly. Regardless of the experience level of the rigger, a safety check is an imperative requirement.

Rope access relies on the concept of double protection – that is, there should always be both an access system to support the worker and a backup system that would protect the worker in the event of failure of the access system, as shown in Figure 8-2. Each of these systems should be completely independent of one another, including anchorages, and should be attached to the technician's harness – although the concept of independent systems does not preclude attachment to the same harness, or even to the same point on the harness.

Before commencing work, all technicians involved should understand the scope and plan of work and the methods that will be used. Access zones and hazard zones

FIGURE 8-2
A Two-Rope System for rope access. Courtesy of Trask Bradbury.

should be identified and marked accordingly, and workers should clearly understand the protective measures that they are expected to use.

An access zone is any area in which personnel may be at risk of falling, such as online or near a working edge. This area should be clearly delineated using warnings, signs, barriers, safety lines, or other methods to prevent or arrest a fall. Where feasible, placing anchorages outside the access zone, in a safe area far enough from any exposed edge, will help facilitate the technician's ability to don PPE and connect to the lines prior to entering the access zone. Placing anchors behind a handrail, parapet wall, or other barrier, as shown in Figure 8-3, is a good choice. Of course, in some work environments anchorages are more appropriately placed directly overhead, in which case adequate precautions must be observed. No person should be permitted to enter an access zone without appropriate protection.

A hazard zone is any area where a person may be at risk (whether by exposure to a fall, falling objects, or work hazards) as a result of the work being performed. Hazard zones can affect both workers and passers-by, and measures should be taken to ensure appropriate protection against anyone being at risk due to such hazards. Where work is undertaken in a public area, hazard zones should be barricaded or marked to prevent entry by unauthorized persons, and in some cases an attendant may need to be stationed near hazard zones to prevent people from wandering into the area unawares.

FIGURE 8-3
The "access zone" is the area where the technician may be at risk of falling; the hazard zone encompasses a wider area in which a worker or bystander may be exposed to any risk as a result of the work. Courtesy of TranSystems Corporation.

Both access zones and hazard zones should be identified and secured, taking into consideration all hazards associated with the type of work to be performed. Where possible, a physical barrier should be preferred, but they should be clearly defined and marked.

8-1 COMPATIBILITY

The term compatibility refers to the appropriateness of any given piece of equipment for the function that it is to perform. Equipment that is designed and intended for life safety use should only be used in a rope access system. Characteristics such as strength, deformation under load, efficiency, and inter-relationships with other equipment are given more emphasis and measured with greater accuracy and statistically higher confidence ratios for life safety equipment than for equipment designed for the commodity market.

Rope access equipment must be selected and used in a manner that is consistent with the use for which it was manufactured, designed, and intended. Employers, supervisors, and technicians should take it upon themselves to confirm that the equipment selected does, indeed, meet the requirements necessary (both performance and regulatory) for the intended application, and that components function appropriately with one another.

8-2 ACCESS SYSTEM

The access system provides the primary support for the rope access technician. The access system consists primarily of an anchored line (rope) onto which descending and ascending equipment are connected. Technicians connect themselves to these via their harnesses, and this facilitates their suspension from and movement along the rope.

The lifeline that will be used for access should be rigged as closely to being in line with the intended work as possible, with offsets as needed to maintain a safe line of travel. This means taking into consideration the fall line, discussed in Chapter 6, and adjusting it as necessary using deviations or re-anchor techniques, described later in this chapter. The work should be planned so as to permit workers to descend and ascend in as straight a vertical line as possible. Every effort should be made to avoid angles that would cause the rope to chafe excessively, anchors to be overloaded, or the technician to be placed in a situation where a pendulum, or swing fall, might occur (Figure 8-4).

When possible, particularly for top-down work, the lifeline should be anchored a sufficient distance back from the edge as described in Chapter 7 to allow the technician(s) adequate working room at the top. This is less critical where work is commencing from the bottom or midline. An access zone should be designated at anchor level and at ground level, and should be of sufficient breadth and position, and/or alternate means of protection provided, to ensure that anyone outside the zone is not at risk of exposure to falling or to other access-related hazards. There should be at least a sufficient amount of area and/or adequate protection to permit the technician to rig onto the rope and begin work from a safe area, with a safe means of access.

If the rope goes over a sharp, abrasive, or hot edge, it should be protected using an edge roller, padding, or other protective material. Other obstructions in the fall line should also be considered for the damage they might potentially cause to ropes

FIGURE 8-4
Access system should be rigged to a secure anchor, as closely as possible to be in line with the work to be performed, and with sufficient protection for the technician to rig into the rope in a safe location. Courtesy of Abseilon USA, www.abseilon.com.

FIGURE 8-5
Where necessary, ropes should be protected. Courtesy of Trask Bradbury.

and equipment, increased potential for entanglement, and avoidance measures that may be required of the technician. These, too, may be padded and protected as necessary, as shown in Figure 8-5.

Where obstructions such as structural components, railings, or parapet walls exist at the edge of the drop, the technicians must ensure that the structural elements are of sufficient strength and design to support the anticipated load. If these elements are not adequate to withstand the effects of the system, the technician may need to adjust the rigging to avoid the obstruction.

Before using any access system for ascent or descent, the rope access technician should terminate the end of the rope with a stopper knot (Chapter 8) to prevent inadvertent descending off the line.

8-3 BACKUP SYSTEM

Whenever a primary rope access system is used for access, the rope access technician should also be attached to a separate and distinct backup system. The foundation of the rope access backup system is a properly anchored vertical lifeline, which is intended to provide protection to the technician in the event of failure of the primary access system. Technicians are connected to the backup system via an appropriate rope access backup device and lanyard, which is in turn attached to their harness – usually via the sternal D-ring.

The backup system should be separate and independent of the primary access system. This includes use of separate anchorage systems so that the failure of any one component should not result in a catastrophic failure of both systems. Of course, reasonable discretion should be exercised here. For example, a single structural anchorage (such as an I-Beam) may be used for attaching both the access and the backup system, as long as that anchorage is unquestionably sound. However, separate anchorage connectors (eye bolts, anchor straps, etc.) should be used for each system (Figure 8-6).

The anchor line used for the backup system should be rigged alongside that of the access system in a manner that permits the technician to be simultaneously attached to, and properly use, both systems without hindrance. A distance of less than 3 ft between the two lines will help facilitate this. Maintaining a short distance between the access line and anchor line will also help prevent the potential for swing

FIGURE 8-6
Backup system should be rigged to be separate and unaffected by the access system – not too close, but not too far apart. A distance of less than 3 ft between systems is advisable. Courtesy of Trask Bradbury.

fall in the event that a fall does occur. These considerations must be balanced with the fact that rigging ropes too closely together can contribute to a risk of entanglement. There are no hard and fast rules to this; it is experience that will offer the best guidance.

As in the case of the access line, the technician should place a stopper knot in the end of the rope used as a backup line before use to help prevent inadvertent travel of the backup device off the end.

When possible, both the access line and the backup line should be anchored in a manner that facilitates nonentry rescue – that is, rescue techniques that preclude the need for a rescuer to expose themselves to a fall hazard should the technician find themselves in need of assistance. This concept is discussed further in Chapter 13.

8-4 ATTACHMENT TO TECHNICIAN'S HARNESS

The technician must be attached to both the primary and the secondary system whenever working at height. Each should be attached with its own independent connector so that any one system may be removed or adjusted at any given time without interference with the other(s). In most cases, the primary system – that which is used for suspension – will be attached to the waist D-ring, while the secondary backup system will be attached to the sternal D-ring. During certain maneuvers, such as rope-to-rope transfer, passing a deviation or when negotiating intermediate anchors, the technician may be connected to more than one primary system, and/or more than one backup system, simultaneously. At such times there may be multiple connections to a single D-ring on the harness, but each should still have its own connector (Figure 8-7).

FIGURE 8-7
Primary access system is typically connected to the waist attachment, while the backup goes to the sternal. Courtesy of Trask Bradbury.

8-5 PULL-THROUGH SYSTEMS

A pull-through system is one wherein the technician can retrieve ropes upon completion of the work, and can do so from a working level that is below the anchorage level. This generally involves some means of rigging the halfway point of a rope to a sturdy beam at a high point, and ensuring that both ends reach the landing zone, so that when the rope is pulled from one side of the anchorage it stays put but when pulled from the other it becomes disconnected. Pull-throughs should be considered as a temporary measure, and are most often used at the end of a job where anchorages and rigging cannot be left in place or at the end of an aid traverse.

There are many variations on the techniques used for creating a retrievable system, but they are generally rooted in one of two concepts:

1. Using a ground anchor with a high point change of direction, and
2. Using a top anchor with some means of a stopper at top.

Pull-Through with Ground Anchor

To use a ground anchor with a high point change of direction, simply anchor the standing end of the rope at ground level by some means (such as a structural element, vehicle, etc.,), pass the working end up through a high point, and ensure that the working end reaches the ground. The working end – that is, the end that is not anchored, becomes the technician's working rope. In this method, keep in mind the force magnification concepts discussed in Chapter 7. Whatever the weight of the technician(s) or load on the working end of the rope, it can effectively be as much as doubled at the anchor, depending on the angle.

When the technician reaches the ground, upon completing the work and disconnecting from the working end, they simply pull on the standing (anchored) end of the line to retrieve the rope.

Top Anchor Pull-Through with Knot

The other approach to creating a retrievable pull-through system involves the use of a midline knot, such as a butterfly, tied into the rope on the standing-end side of the high point anchorage. The working end of the rope is then run through the knot – either directly or via a carabiner – so that the rope tightens around the anchorage when the working end is pulled. As long as the technician uses the working end of the rope for ascending, descending, and/or backup, it remains secure. When the technician reaches the ground they simply pull on the standing end of the line to retrieve the rope (Figure 8-8).

Note that if a pull through approach is being used to create a rope access system, two pull-through ropes will be required: one for the access line and the other for the backup line. This method should be used only with great care by properly trained and experienced technicians, as improper use can result in catastrophic failure, and death (Figure 8-9).

FIGURE 8-8
Pull-through system – knot method. Courtesy of Pigeon Mountain Industries, Inc.

FIGURE 8-9
A pull-through system requires two ropes, an access line and a backup line, as does any rope access system. Courtesy of Pigeon Mountain Industries, Inc.

8-6 CHANGING THE FALL LINE

In some cases, it is necessary for the access and backup systems to be anchored in a location that is vertically out of directional alignment with where the work must be performed. This approach is sometimes used to avoid obstructions in the fall line, or to protect personnel or sensitive equipment that is in the fall line. At other times it may be used simply because terrain features make it necessary.

Where a midline change of direction is necessary, either a directional deviation or a rebelay may be used. Which is the better choice in a given situation will depend on the circumstances. In either case, an anchor must be established at the location where the rope is to be redirected. The strength and nature of this anchor, and how the rope is rigged into it, is determined by the specific performance requirements for the system.

Directional Deviation

The simplest and most straightforward means of redirecting a line is to use a directional deviation, as shown in Figure 8-10. This method involves simply clipping the anchored line into a midline anchor, but not tying it off at the midline point, as discussed in Chapter 7, Terminations and Anchors. In this case, the line is allowed to run freely through an anchored connector or pulley device, which in turn re-directs the rope to a different fall line (Figure 8-11).

The performance requirements of an anchor system for a directional deviation will depend on the intended load, and the probability/consequence of failure. The

FIGURE 8-10
Setting additional anchors midline may be necessary to permit technicians to more accurately position themselves in relation to their work while avoiding swing fall. Courtesy of Burgess and Niple, Inc.

FIGURE 8-11
If a directional deviation is used, both the access line and the backup line must be deviated. Courtesy of Abseilon USA, www.abseilon.com.

force at a directional anchor can reach 1.4 times the force of the load with a 90° bend in the line, but it is generally best to maintain a wide inside angle in a deviation. It is especially important to avoid potentially dangerous pendulum falls. For this reason, a deviation is most appropriate when the offset between the original anchor and the deviation anchor is less than 20 degrees.

Deviated ropes run freely through a carabiner at the deviation, and are not secured with any knots. Because the small change in the path of the rope imparts a low force on the deviation, deviation anchors are often not full-strength anchors. Low-strength deviation anchorages should not be used in situations where a failure of the deviation anchor would be hazardous. When possible, full-strength deviation anchors – or even a short rebelay – should be considered as an alternative. Full-strength anchors help to facilitate the needs of the technician who may need to ascend and descend through the obstruction. Ensuring that the deviation anchor is a full-strength anchor provides greater security and helps to simplify the maneuver.

Rebelay (Re-anchor) Systems

Hazard control is always the dominant factor in rigging for rope access, and as such rigging the deviation to avoid catastrophic potential is of utmost importance. Where a possibility of significant swing fall exists, or where a failure of the deviation anchor could result in a hazardous condition, a rebelay may be the preferred alternative.

A rebelay is a type of system wherein the rope is re-anchored at a midline location, so that the length of the rope above is essentially independent from that of the rope below. This method, also sometimes called a **re-anchor**, also permits multiple technicians to be on a given system, as long as they remain on separately anchored

FIGURE 8-12
Short rebelay (re-anchor). Courtesy of Trask Bradbury.

portions. When re-anchoring a rope, sufficient slack should be left in the top length of the rope to ensure that the upper section of the rope will not become trapped or encumbered by the re-anchor during use. Again, there is no magic formula for this, but an experienced technician will err on the side of leaving too much rope rather than too little. Having too little rope between the offset line and the re-anchor will cause the technician to create side-tension against the re-anchor during transition, thereby increasing both risk and difficulty.

A short rebelay is one means by which technicians can position themselves more closely to their work. As illustrated in Figure 8-12, the original set of ropes that the technician used to get over the edge will result in a work position that is too far from the building wall, due to the overhang. To rectify this, the ropes have been re-anchored beneath the overhang, which will permit the technician to work closer to the wall.

A re-anchor system must be at least as strong as the original anchor, or five times the potential load – whichever is greater. A re-anchor in which the subsequent anchorage is set less than 6 ft offset from the original line is known as a Short Rebelay, whereas a system in which the subsequent anchorage is set more than 6 ft offset from the original line is called a Long Rebelay. The main difference between the two, aside from distance, is the approach that the technician will take in passing the obstruction – which is discussed in Chapter 11 (Figure 8-13).

Well-Being of the Technician

All rope access workers should be properly supervised and be self-supportive. Work teams should consist of at least two people; however, additional personnel may be needed to ensure adequate safety for the workers and for passers-by.

FIGURE 8-13
Long rebelay (re-anchor). Courtesy of TranSystems Corporation.

Technicians require adequate facilities for periodic rest and refreshment, and the frequency of breaks for these purposes should be tailored to the circumstances. The employer and supervisor should consider climatic conditions, difficulty of access, and work type in preplanning breaks, and the supervisor on site should monitor technicians to ensure that the planned breaks are sufficient.

Rope access is not, itself, a particularly grueling endeavor. Any reasonably fit and trained person can learn to work using these methods for access, positioning, and egress. As has been said, rope access is not a job in and of itself, it is merely the bus that one might take to get to a place of work. In the case of rope access, it is not usually the access method that contributes most strongly to how physically taxing the job will be. Rather, it is usually the type of work being performed, the duration of the task, and environmental conditions that impact the worker more profoundly.

Technicians must have ready access to nutrition, refreshment, and fresh water to maintain energy and hydration, as physical exertion, wind, heat, and other factors can have a cumulative effect. Ample opportunity must be given for periodic rest in a location that is protected from the elements. Consideration must also be given to adequate toilet facilities. This last requirement is often a forgotten detail, especially in some of the more remote worksites.

When performing work using rope access techniques, workers may find themselves in situations where they are relatively incapable of adjusting their position to avoid exposure from close proximity to the work being performed. For this reason, certain tools may need to be secured to avoid causing risks to the workers or their suspension equipment.

Small tools may be adequately secured to the worker's harness or any other secure location by means of a lanyard, and/or may be carried in a bucket or bag that is likewise secured to the technician. Heavier tools may need to be suspended separately, on another line. Where tools have connected air hoses, power lines, or other feeds, these may need to be further secured at their upper suspension point and/or at intermediate points to avoid creating undue drag, entanglement, or other hazards. Care should be taken to ensure that any such system does not impair the rope access system or its backup.

8-7 SUMMARY

Safe use of rope access as a part of work at height requires a systematic approach to ensure that personnel are adequately trained and competent, suitable equipment is used properly, and appropriate supervision is provided. A breakdown in any aspect of the system can result in increased risk.

It is the responsibility of the employers and their designee to verify that rope access is an appropriate means of access for the work at hand, and that it is implemented properly.

The concept of double protection is integral to rope access, including both an access system to support the worker and a backup system to protect the worker in the event of failure. Equipment should be used in accordance with the manufacturer's instructions, and rigged using best practices so that the failure of any one component in the system will not result in catastrophic failure.

This systematic approach to work should extend to all technicians, and before commencement of work everyone involved should fully understand the scope and plan as well as the methods to be used. Safety is the responsibility of all those involved in permitting, planning, supervising, and carrying out the work. Ultimately, the primary objective is to organize, plan, and manage rope access work in a manner that minimizes risks and maximizes safety.

CHAPTER 9

Descending

Keith Luscinski

Topics included in this chapter are:

- Descender and rope selection criteria
- Rigging for descent
- Getting on a rope
- Negotiating obstacles and edges
- Managing the descent
- Passing a knot in the ropes
- Rope-to-rope transfer
- Passing a deviation anchor
- Passing a rebelay anchor
- Landing and getting off a rope.

9-1 INTRODUCTION

Descending is the act of moving down a rope system in a controlled manner with the use of a frictional device to facilitate progress along the rope. Although fundamentally similar to sport rappelling, descending differs in that a two-rope system and industrially designed equipment are used and also that it is performed by highly trained technicians. As a matter of semantics, the industrial rope access community avoids the word "rappel" because of its association with recreational caving and rock climbing.

> Although fundamentally similar to sport rappelling, in rope access descending is differentiated by the two-rope system used, highly trained technician and use of industrially designed equipment.

Professional Rope Access: A Guide To Working Safely at Height, First Edition. Loui McCurley.
© 2016 John Wiley & Sons, Inc. Published 2016 by John Wiley & Sons, Inc.

This chapter provides an overview of the equipment, systems, and skills used by an industrial rope access technician to descend a rope system. Rope access technicians spend the majority of their time on descent. Thankfully, descending is one of the physically easiest skills that a rope access technician performs, as gravity does most of the work. However, all of the skills associated with the simple act of sliding down a rope –equipment selection, rigging, working near edges, and so on – require the technician to know more than just how to operate the handle of a descender. This chapter covers all the skills that contribute to a safe and successful descent.

9-2 CHOOSING A DESCENDER

The descender is the workhorse of industrial rope access hardware. It controls what would otherwise be trauma-inducing, free-fall energy and dissipates it as frictional heat along the length of the rope. It is also the most versatile piece of equipment found on a technician's harness. Aside from going down a rope, descenders are used to ascend short distances, haul equipment and casualties, tension rope systems, belay climbers and loads, and serve as load-releasable anchor connections. Many types of descent devices are available, all of which excel at various tasks. Choosing the optimal descender for the job at hand will result in an effortless and controlled descent (Figure 9-1). Choosing a poor descender for the job will provide a frustrating and potentially unsafe, experience.

> The descender is the workhorse of industrial rope access hardware. It controls what would otherwise be trauma-inducing, free-fall energy and dissipates it as frictional heat along the length of the rope.

Manually operated descenders can be classified into the following four broad categories according to their functionality:

1. Descenders with automatic locking and panic locking functions,
2. Descenders with an automatic locking function but no panic lock,
3. Descenders that allow friction modulation but have no locking function, and
4. Descenders with no locking function or friction modulation.

The ability of a descender to lock in place on the rope automatically when released by the technician is undoubtedly the safest and most useful feature of a rope access descender. Devices with this feature will permit movement only when the descent control handle is actuated by the user, preventing an uncontrolled descent in the event of an accidental release of the descender. The automatic locking feature also provides a seamless transition from descent, to two-handed work and back to descent – although most manufacturers recommend a specific handle position or routing of the rope's tail when the descender is intentionally released to perform work. Descenders without an automatic locking function can still be locked in a hands-free position, but require the rope to be securely tied off (usually around the device) to prevent slippage. This extra step, combined with their lack of safety features, makes descenders without an automatic locking function a rare sight on a rope access jobsite.

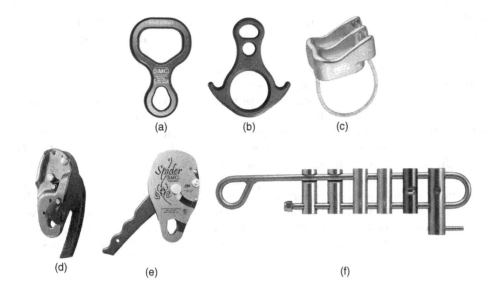

FIGURE 9-1
Various descending devices (a) figure-8, Courtesy of SMC – Seattle Manufacturing Corporation; (b) rescue figure-8, Courtesy of SMC – Seattle Manufacturing Corporation; (c) tube-style belay device, Courtesy of Petzl; (d) automatic locking descender with panic lock, © Petzl; (e) automatic locking descender without panic lock, Courtesy of SMC – Seattle Manufacturing Corporation; and (f) rappel rack, Courtesy of SMC – Seattle Manufacturing Corporation.

Some descenders are also equipped with a panic lock feature, which activates if the technician pulls too hard on the descent control handle. The theory driving these devices is that humans tend to grab on to things tightly when scared suddenly. It's impossible to know how many accidents have been prevented by descenders with this feature, but they surely provide some level of additional safety that goes largely unnoticed in clean, dry, and vertical environments. However, when the force applied to the descent control handle needs to be continuously varied to maintain a constant descent speed, as is the case when ropes get wet and grimy, the panic lock can engage frequently at times when it is not needed. Likewise, some panic locks tend to engage unnecessarily in nonvertical environments, where some of the technician's weight may be on a steep slope rather than on the descender.

Of the four categories of descenders listed, the first two are best suited for rope access work. Descenders of the fourth type (devices without a locking function or the ability to modulate friction) are tube-style or figure-8 belay/rappel devices intended for recreational rock climbing. These types of devices lack the functions and safety features required by a rope access technician. However, stored on the rear of a harness, they may serve as a compact and lightweight emergency descender, in the event of a primary descender malfunction.

Descenders of the third category have a small, but important, niche in the rope access equipment arsenal. Long, continuous descents pose an unusual challenge for the descender due to the self-weight of the hanging rope. At the top of a 500-ft descent, the weight of the free-hanging rope below the technician will cause most descenders to lock onto the rope and refuse to move. Descenders with a panic lock feature tend to perform especially poorly, as the excess rope weight makes the panic lock engage more easily. The technician in this situation will only be able to move downward after lifting the hanging rope to reduce the tension it applies to the descender. Carrying the rope with the technician, as in a bag, is another method of relieving rope weight, although this can be unwieldy and difficult. As the technician progresses downward, the weight of the free-hanging rope decreases and provides less resistance to descent.

Regularly facing such long descents, cavers opt to descend using a rappel rack, which allows friction modulation. A rappel rack can be threaded with only a few friction bars at the top of a long descent and have additional bars added as more friction is needed. Its nonlocking nature, however, means that the rappel rack is not ideally suited for some rope access applications, especially during work requiring frequent stops. Where possible, re-anchoring the rope with a rebelay in the middle of a long, continuous drop can mitigate the effects of rope weight and allow a more conventional rope access descender to be used.

All rope access equipment should be purchased from reputable manufacturers. Depending upon local regulatory bodies and jobsite requirements, certain pieces of equipment may also have to be certified to meet technical standards. Descenders are most commonly certified to the European Standards EN 341 or EN 12841. EN 341 addresses descender devices for rescue[1] while EN 12841addresses rope adjustment devices for rope access systems.[2] At the time of writing, there is no American standard that addresses descenders for use in industrial rope access. However, the American National Standards Institute is working to develop a standard for descent controllers as part of the ANSI Z359 group of fall protection standards. At an international level, ISO 220159 provides test methods and baseline requirements for descenders of all types.

9-3 CHOOSING A ROPE FOR DESCENT

Compared to the descender-selection process, choosing a rope for descending is a relatively simple process. Many descender manufacturers will recommend a specific model of rope to be used with their descenders, but in reality, nearly any life-safety kernmantle rope of a compatible diameter will be suitable. A descender is typically designed and tested to operate only on a relatively narrow range of rope diameters. The rope access industry by-and-large uses 7/16-in. (11 mm) diameter ropes, though ropes of other diameters may be used by specialized organizations, such as fire departments and high angle rescue teams. This standardization across the industry makes it easy to decide on the type of diameter rope to purchase for most scenarios.

Ropes, addressed in greater detail in Chapter 6, can be classified into three categories, according to their elasticity:

1. Dynamic
2. Low Stretch
3. Static.

Dynamic rope is comparatively very elastic and is typically used only in scenarios where a fall is possible, because of its shock absorption properties. The Society of Professional Rope Access Technicians (SPRAT) recommends dynamic rope when a fall of factor greater than 0.25 is possible.[3] Descending, on the other hand, typically has a small fall potential and calls for static or low stretch ropes, as defined by the Cordage Institute.[4] These are discussed in greater detail in Chapter 6.

For many applications that use shorter lengths of rope, static and low stretch ropes are interchangeable without any noticeable effect. However, as the rope length

---

[1] *Personal fall protection equipment. Descender devices for rescue.* European Standard EN 341:2011.
[2] *Personal fall protection equipment. Rope access systems. Rope adjustment devices.* European Standard EN 12841:2006.
[3] *Safe Practices for Rope Access Work.* Society of Professional Rope Access Technicians [SPRAT], 2012.
[4] *Low Stretch and Static Kernmantle Life Safety Rope.* Cordage Institute CI-1801:2007.

is increased, the difference in elasticity between the two types of ropes becomes more apparent. Because the stretch of a rope is proportional to its length, longer descents will feel "bouncier" and are best rigged with static rope.

Where rope lengths are very long, a static rope may also be the safest choice for the backup line because of the rope elasticity's effect on fall arrest distances. Take, for instance, a static rope with 3% stretch at 10% of MBS and a low stretch rope with 8% stretch at 10% of MBS. With the application of 10% of MBS near the end of a 300-ft descent, the static rope will stretch 9 ft and the low stretch rope will stretch 24 ft! That's an extra 15 ft of clearance needed below the technician to allow the backup system to safely arrest a fall (assuming the fall arrest force is equal to 10% of MBS).

9-4 RIGGING FOR DESCENT

With the proper equipment selected, the technician can begin to rig a rope for descent. Anchor selection and construction is the most focused-upon aspect of rigging, but many other small tasks are included in the process of preparing for work on rope. These tasks – such as safely using fall protection near unprotected edges, knotting and lowering rope ends, and assessing weather hazards – are all necessary to safely construct and use a two-rope descent system.

The technician's first step in constructing a descent system is to identify all suitable anchors on the jobsite and construct a mental map of their location. Anchors and anchor systems are addressed in detail in Chapter 8 of this text.

> Each rope should be independently connected to its own anchor capable of holding a 5000-lb load without failing.

With a mental map of all possible anchors, the technician can begin to choose anchors for the intended work location. Ideally, two suitable anchors will be situated side-by-side: one anchor for the work positioning line and one for the backup line. In reality, the anchors may be randomly distributed throughout the jobsite. It is the technician's task to rig the rope systems so that both ropes are adjacent to each other and run parallel to the work location. The chosen anchors should lie outside the access zone if possible, as shown in Figure 9-2, to allow technicians to approach the anchors without fall protection. If the chosen anchors are near a falling hazard, then a separate fall protection system must first be installed to permit safe access to the anchors.

> Anchors should lie outside the access zone to allow technicians to approach the anchors without fall protection.

As part of the process of choosing anchors and corresponding work locations, the technician should be mindful of the weather. Window washers, for example, will want to avoid working on the sunny side of a building in the dead of summer to prevent the windows from drying before they can be squeezeed. On tall structures during a windy day, the leeward side may provide a pleasant work environment while the windward side can be downright hazardous. If work must be performed

FIGURE 9-2
Anchors should be rigged outside the access zone, if possible. Courtesy of Abseilon USA, www.abseilon.com.

in windy conditions, the rope ends can be tied-off or weighted at the landing zone to prevent unwanted rope movement. If potentially foul weather is forecast, the technician should opt to work on an aspect of the structure that provides a view of oncoming weather.

> On tall structures during a windy day, the leeward side may provide a pleasant work environment while the windward side can be downright hazardous.

Once the anchors are rigged, the technician will tie stopper knots at the ends of the descent ropes and *lower* the rope ends to the landing zone. Ropes should only be thrown if it is absolutely necessary to clear an obstacle and only if there is no danger to people or property below (otherwise, the ropes should be bagged and deployed while on descent). Only as much rope as is necessary – plus some extra for wiggle room – should be deployed. A tangle of excess rope in the landing zone only provides unnecessary work for the technician. Keeping excess rope at the top of a drop, rather than below, also helps to facilitate rescue. Because the technician often cannot see the landing zone from the anchors without approaching an unprotected edge, someone with a good view of the landing zone can communicate with them by radio to ensure that both ropes are of adequate length. Without a ground spotter, the technician may need to build a temporary fall protection system to peer over the unprotected edge for a view of the rope ends in the landing zone.

An overview of rigging for descent would not be complete without addressing rescue provisions. Without delving into in-depth discussion on rescue planning, it can simply be stated that the goal of a rope access rescue team is to move an incapacitated technician out of the access zone and into the safe zone, where medical care can be administered. Pre-rigging anchors for hauling or lowering can be one of the fastest and safest methods of achieving this goal in vertical rope systems. In high-risk environments, it would be wise to incorporate a tied-off descender , along with sufficient excess rope at the top to allow for lowering. Alternatively, keeping a fully rigged 3:1 mechanical advantage system ready for hauling can also be a good choice. For everyday applications though, just making anchor connections load-releasable can be a good compromise between speed and safety that eases the transition into

FIGURE 9-3
A handled ascender and foot loop can be used to aid the transition over an edge. Courtesy of Pigeon Mountain Industries, Inc.

rescue operations. A releasable anchor facilitates the transition from descent into rescue operations with very little upfront work.

9-5 GETTING ON ROPE

Before attaching to any rope system, the technician should perform a final safety check of the anchorage components. Anchoring devices should be properly oriented and all textiles should be protected from abrasion. Carabiners should be locked and loaded along the major axis. Knots should be properly tied and dressed with adequate tail. Once the safety check is complete, the technician can attach to the backup line and approach the access zone. As one last part of the safety check before getting on the rope, the technician should look down and confirm that both rope ends reach the landing zone.

Prior to weighting the ropes, the technician will need to prepare rope protection to prevent damage to the rope where it changes direction over the edge. In many situations this rope protection can be as simple as a piece of carpet for padding. For very sharp or hot edges, however, a textile rope protector may be insufficient. Instead, an engineered solution may be required to protect the ropes from especially hazardous edges. All rope protection devices should be secured either to the rope or to the structure to prevent them from becoming displaced and allowing the rope to become damaged.

While standing at the edge, the technician will connect the descender to the primary access line. Both the descender and backup device should be positioned so that neither will be loaded against the edge when the technician transitions from standing to hanging. With the descender locked into the hands-free position, the technician can use both hands to ease over the edge and weight the descender. Some technicians prefer to attach a handled ascender to the work positioning line above the edge, and extend a foot loop from it over the edge, as shown in Figure 9-3, to provide support for one foot while making the transition. Before committing 100% of their weight to the descent ropes, the technician should check the position of the rope protection and make any necessary adjustments. Now over the edge, the technician should inform the rest of their team of his location and "on rope" status.

FIGURE 9-4
Managing descent requires the technician to control the tail of the descent rope, the descender handle (if there is one), and the backup device. Courtesy of Abseilon USA, www.abseilon.com.

9-6 MANAGING THE DESCENT

Once on a rope and clear of the edge, there is little to do but let gravity go to work. Downward progress on a rope is a repetitive process of descending to a work area, stopping to work, then descending to the next work area. Remember that all descenders work by applying friction to the rope, a process that creates heat. While descending, it is best to avoid moving fast enough to overheat the descender and potentially damage the rope. Nylon and polyester fibers melt at 460°F and 480°F, respectively (Cordage Institute), though strength loss occurs at much lower temperatures. Indications that the descender is too hot include: a descender that is too hot to touch, sizzling sounds (especially noticeable on wet ropes), shiny glazing on the rope's sheath and a noticeable smell from the descender. Caution should be exercised when descending long distances without stopping, as even at slow speeds the descender will continue to get hotter. Short distances can be quickly navigated (as is often necessary to descend past a building window without leaving shoe marks), but long descents must be performed slowly and be punctuated with cooling stops.

> All descenders work by applying friction to the rope, a process which creates heat. Short distances can be navigated quickly, but long descents must be performed slowly and be punctuated with cooling stops.

Descending with a handled descender is always a two-handed procedure: one hand operates the descent control handle while the other holds the control rope as it feeds into the descender from below (Figure 9-4). Nonhandled descenders are usually operated with just one hand on the control rope. With all types of devices, the control rope must be managed constantly. While actively descending, a hand must always be kept on the control rope to prevent an uncontrolled descent. During work that requires both hands, the descender must be locked or tied-off according to the manufacturer's instructions before the control rope can be released. Despite the fact that most industrial rope access descenders have an automatic safety locking feature,

FIGURE 9-5
A properly installed rope protecting sleeve. Courtesy of Pigeon Mountain Industries, Inc.

these devices usually have a small extra step that must be performed if the device will intentionally be parked hands-free.

As the technicians work their way down the rope, they will likely encounter additional edges that pose a danger to the descent lines. The most common device used to protect the rope is a canvas or nylon sleeve that can be fastened around both ropes with a Velcro closure. As with going over the initial edge to get on a rope, the rope protector should be prepared before reaching the hazard. Once wrapped around the ropes, the sleeve can be slid down to the appropriate position. The sleeve is then fixed in place with a small diameter cord tied into a friction hitch around one of the ropes, as shown in Figure 9-5. The friction hitch should be tied very tightly and securely to prevent the sleeve from sliding down the rope. A rope protector sleeve that has slid out of place, which often happens when the ropes are pulled up at the end of a descent, is an indication that the friction hitch was not tied securely enough.

Ropes cannot be protected from all hazardous edges and obstacles with a simple rope protecting sleeve. As previously mentioned, some edges are hazardous enough to require an engineered protection system. Another option, when suitable anchors can be located, is to re-route the rope past hazards with the use of a rebelay. However, if an adequate solution to protect the ropes cannot be established, then the descent cannot be performed. General abrasion along the length of the rope is to be expected as the rope ages, but severe and acute damage to the rope is not normal wear; it is caused by technician negligence.

> General abrasion along the length of the rope is to be expected as the rope ages. Severe, acute damage to the rope is not normal wear; it is caused by technician negligence.

9-7 TENDING THE BACKUP DEVICE

Although many different models of backup devices are used in the rope access industry, they can all generally be classified into two groups: those that must be manually moved down the rope, and those that automatically follow the technician down the rope. Regardless of the type, all backup devices should be kept high to minimize potential fall distance in the case of a work positioning line failure.

Many manually operated backup devices can be defeated if held or grabbed by the technician in the event of a fall. The same action that is intentionally used to move the device down the rope can also delay the device from engaging during a fall. Ideally, when using these backup devices – *especially* when paired with a descender without a panic lock – the technician should move them independently of the descender. However, the inclusion of "tow-strings" with many of these devices has made it common practice for technicians to descend while simultaneously towing the backup device. It would be preferable for the technician to descend a couple of feet, then stop and reposition the backup device.

Self-trailing devices are convenient in that they usually don't require tending during the descent. However, the feature that allows these devices to follow the technician down the rope can also cause poor behavior in windy conditions. On long descents, the wind loading on the rope above the technician can pull the rope through these devices and create an arc of slack between the technician and the overhead anchor. Manually operated backup devices in this scenario will maintain their position on the rope and prevent excess slack from generating above the device. When self-trailing devices are used in windy conditions, it is often helpful to weight the tails of the backup rope to reduce rope deflection caused by wind loading – especially when using a self-trailing backup device (Figure 9-6).

9-8 PASSING A KNOT

Knots are tied in every work positioning and backup line for two primary reasons: to connect the rope to an anchor and to prevent a descent off the end of the rope. Once on descent though, the technician typically doesn't interact with knots in the work

FIGURE 9-6
In windy conditions, it is often helpful to weight the backup line. Courtesy of Trask Bradbury.

positioning or backup lines. However, if the middle of a rope is severely damaged during the course of work, a butterfly knot can be used to temporarily isolate the damaged section and allow work to continue. Short ropes can also be tied together in a pinch (using knots such as the Flemish bend or double fisherman's bend) to effectively create one long rope. In either scenario, the technician will have to pass the knot to continue descending.

Remember that the ability to temporarily isolate a damaged section of rope with a butterfly knot is no excuse for poor rope protection practices. Before tying a butterfly knot and continuing the work, the first priority is to identify and rectify the cause of the rope damage. On that note, tying short ropes together to achieve a sufficient length rope is a sign of ill-preparedness. The best course of action is to arrive on site with the proper length ropes. The task of passing a knot, though simple, can often be easily avoided through good practice and preparation.

Passing a knot in the backup rope while descending is indeed a simple task, especially for the technician equipped with two backup devices:

1. Descend until the knot is at chest level.
2. Attach the second backup device directly below the knot.
3. Remove the first backup device.

The key to a safe and efficient knot pass is to descend close to the knot without jamming the backup device on top of the knot. Stopping before the knot reaches chest height will result in a second backup device that is too low once it is positioned below the knot.

If the technician is equipped with only a single backup device, he is tasked with the additional step of creating a temporary connection to the backup rope with a knot and carabiner:

1. Descend until the knot is at chest level.
2. Tie a figure 8 on a bight knot about 1 ft *below* the knot to be passed. The short length of the rope between the two knots must be wide enough to fit the backup device.
3. Connect to the figure 8 knot directly to the sternal D-ring with a locking carabiner.
4. Remove the backup device from the rope and reinstall it below the knot to be passed, but above the figure 8 knot.
5. Disconnect and untie the figure 8 knot.

The placement of the figure 8 knot below the backup device is an important step to avoid generating excess slack above the backup device. If the backup device were instead reinstalled below the figure 8 knot, it could move dangerously low when the figure 8 knot is untied. Also realize that if a butterfly knot is in the backup line to isolate a damaged section of the rope, that loop cannot be used as the temporary backup connection while the backup device is moved.

Passing a knot in the work positioning line while descending is slightly more complicated, and requires a brief use of the ascenders to unweight the descender and transfer it below the knot:

1. Descend until the knot is several inches from the descender.
2. Changeover onto the ascenders.

3. Remove the descender from the rope and reinstall it directly below the knot.
4. Descend on the ascenders until the knot is only an inch or two from the lower ascender.
5. Changeover back onto the descender.
6. Remove the now unweighted ascenders.

Passing a knot, shown in Figure 9-7, is really not difficult; it simply requires the technician to perform a standard changeover from ascent to descent. However, unlike a standard changeover, which can be accomplished with very little drop in the technician's height from start to finish, performing a changeover to pass a knot necessitates a drop in the technician's height. This transition can be eased by positioning the ascenders and the descenders as close together as possible (steps 3 and 4), separated only by the length of the knot. Keeping the ascender with the foot loop low will also prevent finishing the changeover with the foot in an uncomfortably high position. To ensure a minimal drop in height, the ascenders and descenders should be as close together as possible, separated only by the length of the knot.

Remember that practicing new maneuvers on a rope will always feel unnatural at first. Running through the steps several times will help smooth things out. As long as connections to both the work positioning and backup ropes are maintained, the maneuver, no matter how rocky, can be deemed a success.

FIGURE 9-7
Passing a knot in the backup line. Vertical Rescue Solutions by Pigeon Mountain Industries, Inc.

9-9 PASSING A DEVIATION ANCHOR

The ability to move laterally while on the rope is often paramount to the efficiency of rope access work. Every field of rope access requires it. From window washing to offshore welding, technicians need to move around to get work done. In many situations, it's as easy as pushing against the structure with one's feet to move sideways. However, for descents down blank vertical walls, such as dams, and down free-hanging ropes, a technician has nothing to assist out-of-plumb movement. In these situations, it can be convenient to run the ropes through an anchor that redirects the path of the rope overhead.

Deviations, discussed in Chapter 8, are anchor points used to pull the plumb line of the ropes from the main anchors into a more convenient location and are used to move the ropes only a short distance.

The successful navigation of a deviation begins during rigging. A poorly rigged deviation, with a sling that is too short or that has no provision to clip a lanyard, will frustrate even the best technicians. The first step is to add a sling to the anchor point that is slightly longer than a harness lanyard. The work positioning and backup ropes will run through a single locking carabiner at the end of this sling. A single locking carabiner is also added directly to the deviation anchor. A large knot should be tied in both ropes about 15 ft below the deviation to allow the technician to pull over to the deviation on descent (Figure 9-8).

To pass the deviation on descent, the technician descends until eye-level with the deviation, then uses the ropes to pull horizontally to the anchor. Once connected to the deviation anchor with a lanyard, the deviation sling is repositioned above the descender and backup device.

Step by step, this maneuver should go as follows:

1. Descend until eye-level with the deviation anchor.
2. Pull on both the work positioning and backup ropes until the large knot jams in the deviation carabiner. Pull toward the deviation anchor with a hand-over-hand motion.

FIGURE 9-8
Passing a Deviation on descent. Vertical Rescue Solutions by Pigeon Mountain Industries, Inc.

3. Continue pulling sideways until a lanyard can be clipped from the harness to the carabiner directly on the anchor (not the deviation carabiner at the end of the sling).
4. While held in place laterally by the lanyard, unclip the deviation sling from the work positioning and backup ropes.
5. Re-clip the deviation sling to both ropes above the descender and backup device. If the sling isn't long enough to be re-clipped above the descender, it may be necessary to descend a few inches farther.
6. Pull toward the deviation anchor just enough to unclip the lanyard.
7. Slowly lower out with the hand-over-hand motion until the deviation sling is under tension.

The most common source of difficulty in passing a deviation is poor rigging practices. Pulling laterally toward the deviation should require muscle, but should not be strenuous. Such difficulty is a sign that the deviation angle is too large and that the anchor point must be moved closer to plumb with the main anchor. Where a deviation cannot be made less strenuous, a rebelay should be employed. Difficulty in re-clipping the deviation sling above the descender may be easily addressed by descending a few inches farther, as mentioned above. However, if the deviation sling cannot be easily re-clipped in any position, it is likely that it is too short and should be extended.

As a final note on deviations, it should be noted that the deviation anchor is often not a full-strength anchor. During the process of passing a deviation, the anchor is clipped only as a temporary positioning aid. Neither the backup device nor descender should be removed at any point while descending past a deviation anchor.

9-10 PASSING A REBELAY

A rebelay is similar to a deviation, but with two important distinctions. The first is that the ropes are fixed to the rebelay anchors with knots rather than loosely run through a carabiner. When the technician is below the rebelay, all of his or her weight will be on the rebelay anchor and the main anchor will be unweighted. This allows a permanent sag in the rope to be rigged into the system between the main and rebelay anchors. As is discussed later, this hanging loop in the rope is necessary for moving through the rebelay system. The second distinguishing feature is that the rebelay point comprises two full-strength anchors (one for the work positioning line and one for the backup line), where the deviation point may employ only a single, non-full-strength anchor for both ropes. This makes a rebelay anchor different from a main anchor only in its location in the rope system.

The horizontal distance between the main anchor and rebelay anchor has no limitation, as it does with the deviation. Rebelays tend to be grouped into two categories, depending on this horizontal offset. The short rebelay, with its roots in the vertical caving world, is often used in systems with zero horizontal offset. Facing shafts hundreds of feet deep, cavers often choose to re-anchor the rope at intervals rather than have one long continuous length of rope to ascend and descend (Alpine Caving Techniques). This use of the short rebelay on long drops helps to protect ropes in the harsh cave environment.

In rope access, a short rebelay mitigates the effects of the elasticity of a continuous length of rope that is hundreds of feet long. It also distributes the weight of the rope among all of the rebelay anchors, which can be unwieldy at the top of a long

PASSING A REBELAY

FIGURE 9-9
Long rebelays are used to allow the technician to traverse horizontal distances. Courtesy of TranSystems Corporation.

descent. When many people are required to move up or down a long rope, this technique can also greatly aid efficiency by allowing several people to be on the rope at a time, each on a separate and independently anchored section of the rope.

Conversely, long rebelays are used as a method for horizontal movement without dragging the ropes below. A rebelay is considered to be a long rebelay when the rebelay anchor is greater than 15° from plumb with the main anchor (note that an offset of this magnitude would preclude the use of a deviation). Long rebelays are indicated where the plumb line of the rope must be moved more than a few feet (Figure 9-9).

Aside from their geometry, short and long rebelays are rigged with identical principles: full-strength rebelay anchors and sagging rope loops between the main and rebelay anchors. Similar techniques are used to negotiate both long and short rebelays on descent, though very wide rebelays can require extra effort.

To pass a long rebelay on descent, the technician uses changeovers to progress through three different sets of ropes, as shown in Figure 9-10. While transitioning through the loop of rope between the main anchor and rebelay anchor, the descender, ascenders and two backup devices are used to accommodate the changeovers.

One method for passing a long rebelay is as follows:

1. Descend from the main anchor until eye-level with the rebelay anchor.
2. Attach ascenders to the far side of the rebelay loop, as close to the rebelay anchor as possible and oriented to load the rebelay anchor.
3. Attach a second backup device to the far side of the rebelay loop, with the same orientation as the ascenders. If not equipped with a second backup device, tie a figure 8 knot as close to the rebelay anchor as possible and connect it with a lanyard from the sternal D-ring.
4. Descend until plumb below the rebelay anchor, and the ascenders are fully weighted and the descender is fully unweighted. On long rebelays, it may be necessary to move laterally by climbing on the ascenders and lowering on the descender until plumb below the rebelay anchor.
5. With the descender unweighted, remove it from the rope. Also remove the original backup device from the rope. The technician should now be hanging plumb below the rebelay anchor, connected to the loop of rope between the main and rebelay anchors.

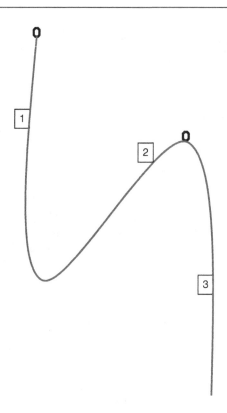

FIGURE 9-10
To pass a long rebelay on descent, the technician uses changeovers to progress through three different sets of ropes.

6. Transfer to the ropes hanging plumb below the rebelay anchor by reattaching the descender and original backup device to those ropes.
7. Changeover from the ascenders to the descender.
8. Remove the ascenders and corresponding backup device or backup knot.

Using both sides of the loop for movement allows the technician to move through the loop without dragging the bottoms of the horizontally rigged rope(s) across the span.

In negotiating a short rebelay, one of these changeovers may be eliminated if desired. For this maneuver, the technician may transition directly from the initial descent line to the ropes hanging plumb beneath the rebelay anchor, as shown in Figure 9-11.

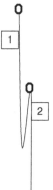

FIGURE 9-11
To pass a short rebelay on descent, the technician need only negotiate two different sets of ropes.

One method for passing a short rebelay is as follows:

1. Descend on the first set of lines from the main anchor until eye-level with the rebelay anchor.
2. Attach ascending system to the primary rope hanging plumb below the rebelay anchor, oriented to load the rebelay anchor.
3. Attach a second backup system to the backup rope as it hangs plumb below the rebelay anchor, also oriented to load the rebelay anchor.
4. Descend on the first set of lines until the ascending system is loaded on the rope below the rebelay anchor.
5. With the descender unweighted, remove it from the rope. Also remove the original backup device from the rope. The technician should now be hanging plumb below the rebelay anchor, connected to the ropes that extend straight down.
6. Either descend on ascenders, or changeover from the ascenders to the descender.

9-11 LANDING

Once the work is complete and all obstacles are passed, the technician can descend to the landing zone and get off the rope. However, before reaching the ground, the technician should be mindful of the hazard of working on the rope near the ground or other obstacles. As discussed in the section on rope selection, a long descent coupled with an elastic rope can mean that the backup system will be unable to prevent a ground impact if the working line fails when the technician is 20 or even 30 ft above the ground. The technician should always be aware of the clearance distances required by the backup system and avoid lingering in locations where the system may be incapable of safely arresting a fall.

> A long descent coupled with a very elastic rope can mean the backup system will be unable to prevent a ground impact if the working line fails when the technician is 20 or even 30 ft above the ground.

The technician should also ensure that their intended landing zone is an area that does not require fall protection. Just as it is preferable to select anchors outside the access zone, it is preferable for the descent to finish in an area without fall hazards. Once on a walkable surface, the technician may opt to disconnect from the work positioning line, but not until the technician is clear of all fall hazards should he disconnect from the backup system.

Landing on the ground plumb below the descent ropes and with feet shoulder-width apart will help the technician stay balanced as he transitions from hanging to standing. As the descent ropes are unweighted, the work positioning line will recoil back to its unstretched length. On long descents requiring the use of rope protecting sleeves, this recoiling action will often move the sleeves out of place and leave the bare rope unprotected. During work near especially hazardous objects, the descent lines should not be accessed again from the bottom (e.g., to ascend) once they have been unweighted, unless it is certain that all of the rope protecting sleeves are in place. If the ropes are continuously weighted and unweighted, the best option may be to attach rope protecting devices to the structure, rather than the ropes.

After the technician is clear of all fall hazards and disconnected from both ropes, the team can be informed that they are "off rope." This not only informs the other team members that they are safely on the ground, but that their ropes and rigging may be moved to the next work location. If the technician has completed the work for that descent location, the stopper knots should be untied from the end of the rope to prevent snagging when the ropes are pulled up. In windy environments or populated areas, the technician should remain in the landing zone to control the rope ends until they are hoisted up.

9-12 SUMMARY

Descending is a fundamental and necessary skill for any rope access technician. Mastering good and proficient techniques is possible only through experience and committed effort.

Although recreational rock climbing may help provide an idea of whether a person has the appropriate aptitude for work at height, this experience is far from sufficient in preparing a person for descending techniques in the workplace. Today's professional rope access equipment, systems, and techniques are all quite different from their recreational counterparts, and require special training and knowledge to be used to their full potential.

Descending is just the beginning of appropriate skills. The rope access technician who descends must also be capable of extricating themselves from a stuck predicament, as well as be capable of ascending out of any situation into which they might get themselves. Adhering to this tenet helps to ensure the continued stellar safety record of rope access, and helps keep the rope access technician safe.

CHAPTER 10

Ascending

By the end of this chapter, you should expect to understand:

- Equipment for ascending
- Standard rigging for ascent
- Alternative ascending systems
- Getting off the ground
- Managing ascending equipment
- Tending the backup device
- Passing a knot on ascent
- Passing a deviation on ascent
- Passing a re-anchor (Long, Short) on ascent
- Negotiating an edge on ascent,

Ascending a rope is considered a fundamental skill for any rope access technician, even one who is at the most basic level. It is at least equal in importance to descending, and some would argue that it is even more essential.

While ascending is more physically taxing than descending, with good technique and a little practice a rope access technicians can use their skills to become more efficient in their work, capable of reaching more challenging locations, and be safer in that they will be more likely to be able to get themselves out of any predicament that they might get themselves into by descent.

As with descending, the technician should always use a two-rope system when ascending, as shown in Figure 10-1, with one rope as the access (ascent) line and the other for backup safety. An important part of this skill is the ability to transition from ascent to descent, and back again from descent to ascent, maintaining connection to both systems at all times, while fully suspended by the rope.

This chapter will cover the essential skills that contribute to ascending safely and effectively.

Professional Rope Access: A Guide To Working Safely at Height, First Edition. Loui McCurley.
© 2016 John Wiley & Sons, Inc. Published 2016 by John Wiley & Sons, Inc.

FIGURE 10-1
Always ascend using a two-rope system, with one rope for access and the other for backup. Courtesy of Abseilon USA, www.abseilon.com.

10-1 SELECTING ASCENDERS

Ascenders used for rope access are typically mechanical devices that incorporate a camming mechanism that permits movement along the rope in one direction, but grips the rope when pulled in the opposite direction.

There are two types of ascenders commonly used in rope access, the handled ascender and the chest ascender. As described in Chapter 6, the handled ascender features a built-in hand-grip for the technician to grasp while ascending while the chest ascender is specially designed to mount on the technician's harness. These two are used together to create a complete ascending system, described in more detail in the following text.

Ascenders perform a function similar to, but should not be confused with, rope grabs that are designed for backup safety or for general rigging use. While certainly robust enough for personal ascending, this type of rope grab is not designed for this purpose and can therefore be more challenging to use. See Chapters 5 and 6 for more discussion on ascenders.

Handled Ascenders

Handled ascenders should fit comfortably in the technician's hand and should be easy to get on and off the rope, yet not prone to accidental disengagement. They should grip the rope well, and be strong enough to hold up under industrial use. Handled ascenders, especially those made for recreational climbing, are usually made of lightweight aluminum alloys. The technicians should choose an ascender that will not bend too easily under their working conditions. In addition, some ascenders feature cams with quite aggressive teeth. The technicians should choose their ascenders carefully to ensure that they will not damage the rope, and that they meet their particular needs. A handled ascender should never be used for fall arrest.

Chest Ascender

Chest ascenders are quite similar to handled ascenders except that they are shorter in length, do not have a grip-handle, and are designed with a twist in the frame and strategically placed holes to facilitate connecting between the waist and sternal attachments on the technician's harness, at about the level of the xiphoid process. The twist in the frame helps the chest ascender to maintain flatter alignment against the technician's body.

The chest ascender and handled ascender must be used together as part of a complete system, as illustrated in Figure 10-2, to be considered an appropriate means of suspension for the technician. The system should be used in a manner that limits potential forces to less than 900 lb on the equipment, or the maximum force recommended by the manufacturer, whichever is lower.

Ascenders must be placed on the rope in the appropriate orientation for the desired direction of travel. When the technician slides the ascender up the rope, toward the direction of travel, the ascender should slide easily. When the technician pulls the descender downward, opposite to the direction of travel, it should clamp down on the rope firmly and prevent movement in that direction.

Both chest ascenders and handled ascenders are designed to be operated by the technician with just one hand. The cam should be designed with a holding mechanism that will permit its being temporarily held open without user intervention. Most ascenders feature a small toggle that may be used for this purpose. Once clamped onto the rope the ascender should not be able to come off the rope without deliberate action by the technician.

10-2 THE COMPLETE ASCENDING SYSTEM

There are multiple ways to rig a safe and effective ascending system, but this text will address only the setup most commonly used in rope access. This system strikes a good balance between efficiency in climbing and ease of transitioning from ascent to descent (and vice-versa). Technicians who may spend most of their time climbing, and relatively little time descending or transitioning, may elect to use different systems. Regardless of the type used, the system should meet the following criteria:

1. Reduces potential fall distances
2. Reduces potential impact forces
3. Maximizes the return on effort expended by the technician
4. Permits the technician to change over from ascent to descent if needed
5. Accommodates self-rescue
6. Accommodates partner-rescue.

FIGURE 10-2
Technician equipped with handled ascender and chest ascender. Courtesy of Vertical Rescue Solutions by Pigeon Mountain Industries, Inc.

The system most commonly used in rope access requires:

(a) a Harness with chest ascender attached, and
(b) a Handled ascender with cow's tail and footloop attached.

Of course, any access system should also be used with an appropriate rope access backup system.

Rigging the Chest Ascender

The chest ascender must be firmly affixed to the users' harness, as they will rely upon it with total commitment during ascent. The precise manner of attachment will depend largely upon the design of the harness, but there are a few key concepts that facilitate safe and effective connection.

Any harness used for rope access should have connection points at both the sternal and waist locations. The sternal attachment of the harness is generally a D-ring that is most often used for connecting the secondary backup safety system, while the

THE COMPLETE ASCENDING SYSTEM

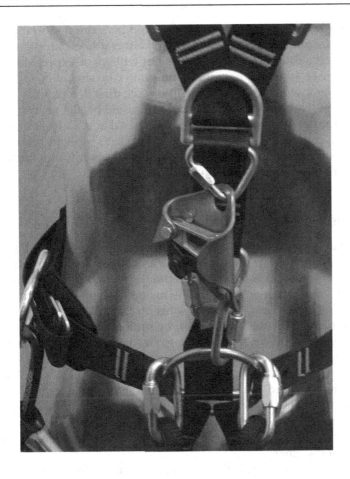

FIGURE 10-3
Chest Ascender attachment to full body harness. Courtesy of Vertical Rescue Solutions by Pigeon Mountain Industries, Inc.

waist attachment D-ring is the primary point of connection for the descender and other access equipment. The chest ascender is actually most effective when placed comfortably between the two. It should be positioned low enough to ensure that most of the load of the technician's body is taken by the seat portion of the harness, but high enough to stay out of the way of the descender during changeovers, and to help keep the wearer in a relatively upright position when using this as a point of suspension during work.

Many harnesses are designed with the intent that the primary attachment of the chest ascender is directly to the D-ring at the waist, with an upper connection to the sternal D-ring for security (Figure 10-3). However, some harnesses have separate connection points just for the chest ascender, and some even feature a permanently affixed chest ascender built into the harness. It is the lower attachment that takes the majority of the weight of the user, while the upper attachment is largely to help keep the ascender properly positioned, and the wearer upright.

The chest ascender should be attached only to manufacturer-approved locations and in an approved manner. In most cases, the chest ascender is best attached to the waist D-ring with a screw link connector. It should be ensured that only screw links that are specifically approved for life-safety purposes are used. The compact dimensions of the screw link are preferred over other types of connectors in this position, as this helps keep the chest ascender properly aligned and helps to avoid cross-loading. The chest ascender should be positioned comfortably against the wearer so that the open portion with the cam faces the outward direction, where it can be easily connected to the host rope, and so that when it is in use the cam accommodates travel in an upward direction.

The upper hole in the chest ascender should be connected to the chest D-ring or other manufacturer-approved location with either a screw link or an appropriate soft link. This connection should be secure, but will not be expected to take the full weight of the user. When making this connection care should be taken to not pull the chest D-ring downward or in an inappropriate direction of pull.

When properly rigged, the chest ascender should be aligned comfortably, held in position (but not too tightly) between the sternal and waist attachments of the harness, at approximately the level of the technician's xiphoid process. When the wearer inserts a lifeline into the ascender and sits down in the harness, the cam should prevent them from sliding downward on the rope.

Rigging the Handled Ascender

The primary purpose of the handled ascender is to grip the access line while holding a footloop, which the technician will use as a step to make upward progress on the rope. If the footloop is of a single loop design, it should be adjustable so that the technician can shorten or lengthen it as needed. Alternatively, a multistep etrier (or soft webbing ladder) may be used in this position. Regardless of which type of footloop or etrier is used, the length should permit the technician's foot to rest comfortably in it with his/her knee bent and leg raised when the technician reaches with the ascender to an arm's length above the head.

For safety, the handled ascender must also be securely attached to the technician with a cow's tail (lanyard) that is long enough to permit the technician to raise the handled ascender about an arm's length above him on the rope yet still be able to operate the cam with one hand. The cow's tail is connected at one end to the handled ascender (with a locking connector, as it is a life safety connection) and to the user's waist D-ring at the other.

The cow's tail connection to the user should be made in as compact a manner as possible, either directly with a knot or with a screw link. Again, the large dimensions of a carabiner render that type of connection undesirable, due to the bulk and the potential for cross-loading.

10-3 MANAGING THE ASCENT

Technicians who wish to ascend a rope from the ground up will first attach themselves to their system as follows:

1. Attach backup device to the safety rope, and push it as high as it will comfortably reach on the rope without the technician leaving the ground.
2. Place access rope into the chest ascender, removing all slack and pushing it as high as it will go.
3. Place handled ascender onto access rope, above chest ascender.
4. Ensure that all connectors are locked.
5. Sit comfortably in seat harness.

From here, beginning the ascent is a relatively simple procedure:

1. Place foot into the footloop and press downward, simultaneously sliding the chest upward along the rope. If there is little rope weight below, it may be necessary to provide back tension on the rope below the chest ascender by

MANAGING THE ASCENT 185

pulling downward on the rope below the chest ascender with a hand, or by "trapping" the rope with the feet, so that the chest ascender slides upward.
2. Sit down into the harness.
3. Relax tension on the leg and push handled ascender as high as it will comfortably reach.
4. Stand up in footloop, simultaneously pushing the chest ascender as high as it will go on the rope as in Step #1.
5. Repeat Steps 2–4 until the desired height is reached.

Ascending rope, as shown in Figure 10-4, requires good physical fitness. Keeping the body as vertical as possible and maximizing the use of the legs will help minimize fatigue.

It is also possible to use the ascending system to downclimb the rope. Downclimbing with an ascending system is more tedious than descending using a

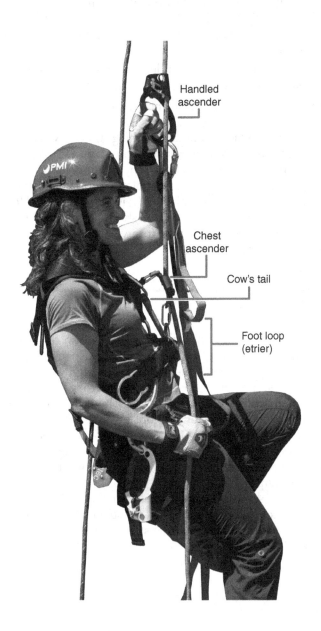

FIGURE 10-4
Good body position and technique contribute to ease in ascending rope. Vertical Rescue Solutions by Pigeon Mountain Industries, Inc.

descender (Chapter 9), but if the technician is descending only a short distance this method may be more efficient than doing a changeover (described later in this chapter).

A technician who is suspended in an ascending system may downclimb using the following procedure:

1. Sit comfortably in harness.
2. With foot in footloop, move the handled ascender down on the rope so that it is just high enough to permit a short step up.
3. Stand up in footloop so that the tension on the chest ascender is released.
4. With the tip of one finger, press downward on the cam of the chest ascender so that it is not cammed onto the rope – *but do not open the mechanism that would completely release it from the rope*.
5. Squat slightly so that the chest ascender moves down on the rope to a point lower than where it began. Move in small increments, so that each downward step does not apply tension to the handled ascender. Moving too far in one move will result in being unable to release the handled ascender for the next downward step.
6. Release the cam of the chest ascender so that it once again grips the rope.
7. Sit comfortably in the harness.
8. Repeat Steps 2–7 until the desired height is reached.

When the technician reaches the desired height, they will either remain there, suspended, to perform work, or will transition off of the ascending system and onto an appropriate platform, or will transition from ascent to descent.

When using an ascending system, the potential for a significant fall or impact force should be mitigated. Ascenders are not designed or intended to take significant impact forces or shock loads, so this potential should be avoided when using ascenders. Descenders are generally much more forgiving than ascenders where potential for impact force is concerned.

10-4 CHANGEOVERS

At times it becomes necessary to change over from an ascending system to a descending system (or vice-versa) while remaining suspended from the rope. The term "changeover" simply refers to the process of changing from ascent to descent, or from descent to ascent, while suspended on a rope. This technique may be readily accomplished with a little practice, and should always be performed in a manner that maintains two points of constant positive connection at all times. The technician must always be connected to at least one access system and at least one backup system. In some situations, such as during a long rebelay, the technician may actually be connected to four points of positive connection: two primary access systems and two backup systems.

If changing from descent to ascent, the technician uses the handled ascender and etrier to step up and connect the chest ascender above the descender, then removes the descender from the line. Step by step instructions for changing from descent to ascent is described in Chapter 9.

Changeover from Ascending System to Descending System

If changing from ascent to descent, the technician simply places the descender on the rope below the chest ascender, locks it off, then uses the handled ascender and etrier to step up, unweight the chest ascender, and remove it from the rope, then sits back down into the descender. To change from ascent to descent, the following steps may be used:

(a) Begin by sitting comfortably in the harness, suspended by the chest ascender and with the handled ascender attached to the main line above the chest ascender in typical ascent configuration.

(b) Connect the descender to the main line, just below but as close as possible to the chest ascender.

(c) Lock off the descender in accordance with manufacturer's instructions.

(d) Adjust the position of the handled ascender so that it is just above head level. The goal here is to ensure that when the chest ascender is removed and the technician is suspended from the descender, the ascender will still be within reach.

(e) Adjust the position of the backup device so that it is just above the sternal attachment. The goal here is to ensure that it will not be inadvertently tensioned and/or out of reach after the changeover.

(f) Using the foot loop, step up to release tension from the chest ascender.

(g) Open the chest ascender and remove it from the main line.

(h) Sit comfortably in the harness, now suspended from the descender.

(i) Remove the handled ascender from the main line and stow it properly.

The ability to change over from ascent to descent, as described above, and from descent to ascent, as described in Chapter 9, forms the foundation for virtually all the additional maneuvers performed regularly by the rope access technician. The technician should be capable of performing these skills smoothly and proficiently before moving on to other maneuvers.

Using a Descender for Ascent

Most autolocking descenders may also be used to ascend the rope, instead of the chest ascender. While not being the most efficient method of ascending, it is often advantageous to use this technique for short distances and when passing obstructions.

When using an autolocking descender for ascent, the technician should set his system up as shown in Figure 10-5. The descender is connected to the technician's waist attachment in the usual manner, and the access line reeved through it in a normal fashion. The handled ascender with footloop and cow's tail are rigged in the usual manner.

To ascend using a descender connected to the waist attachment, the following steps are used:

1. Sit comfortably in the seat harness, supported by the descender.
2. Place foot into footloop.

FIGURE 10-5
Technician rigged to ascend with descender. Vertical Rescue Solutions by Pigeon Mountain Industries, Inc.

3. Bend knee and push the handled ascender as high as it will comfortably reach.
4. Unlock the descender while maintaining control of the slack end of the primary access line.
5. Stand up in footloop, simultaneously pulling as much slack as possible through the descender.
6. Lock off descender or maintain control of the slack end of the line.
7. Sit back into seat harness, weight on the descender.
8. Repeat Steps 2–5 until desired height is reached.

This method can be particularly taxing on the arms, and is therefore most commonly used only over short distances.

Rope-to-Rope Transfer

The basic skills of ascending, descending, and performing a changeover provide the foundation for transferring from one set of ropes to another. Rope-to-rope transfers are used when the technician is using multiple sets of rigged lines to move in a horizontal direction across a structure, or whenever it becomes necessary to transfer from one set of ropes to another.

In most cases, whether the ultimate goal of the technician is to progress upward or downward on the system, the rope-to-rope transfer is best facilitated if the technician begins the process with the descending system attached to the lines they are departing, and the ascending system attached to the lines that they are progressing toward. This means that if the technician initially is in ascent mode, to perform a rope-to-rope transfer they should first change over from ascent to descent on the set of lines they are departing.

If the distance is reasonably short, the technician may simply ascend on his descender to the desired height, then perform the transfer. However, often the technician may find it less grueling to ascend in his ascent system and perform a changeover from ascent to descent on the first set of lines prior to making the transfer.

To perform a changeover, the technicians must ensure that they are first able to reach the second set of ropes – the ropes to which they intend to move – without creating a pendulum or swing fall potential. In most cases, this means collecting the second set of ropes before beginning ascent or descent on the first set, and carrying these until they are needed. This may be accomplished simply by securing the second set of ropes to the harness with a connector and trailing them during the course of work.

To perform a changeover from ascent, proceed as follows:

(a) Prepare to ascend the first set of ropes, with the second set of ropes secured within reach for later use.
(b) Ascend to the desired height and perform a changeover to descent if necessary.
(c) Lock off the descender in accordance with manufacturer's instructions.
(d) Connect the ascending system to the second access line, as high as comfortably feasible, but without pulling away from the fall line.
(e) Connect the backup system to the second backup line, just above head level.
(f) Descend back down the first set of lines until the ascending system becomes taut on the second set.
(g) If necessary, alternately ascend on the second set of lines and descend on the first set of lines to adjust and maintain appropriate position until suspended vertically on the second set of lines.
(h) Use the two sets of lines to carefully control horizontal movement and avoid swing fall potential. Maintain an appropriate angle (usually less than 120°) between the two ropes so as to keep anchor forces at an acceptable level.
(i) Once suspended from the second set of lines and the ascending system, remove the descending system (and backup) from the first set of lines.
(j) Ascend or descend the second set of lines, as desired.

Passing a Knot in the Ropes While on Ascent

Occasionally, a technician may ascend to a certain point in the rope only to find knots obstructing the path. Knots may exist in ascent lines (and corresponding backup lines) where two ropes are joined together, where a damaged section of rope has been isolated, or for other reasons. Knots in an ascent or backup line are not a significant challenge, but must be approached systematically to maintain adequate connection and redundancy at all times.

If knots exist in both the primary access line and in the backup line, it is especially important to ensure that the backup safety system is managed in such a way as to ensure that the backup device stays sufficiently high. This means that it should always be kept at a level on the backup line that is higher than the height of the technician's sternal attachment. When possible, it is desirable to pass the knot first on the backup system, and then on the access system.

To pass a knot in the backup line, follow these steps:

1. Ascend as high as possible on the primary line, while keeping the backup device at sternal attachment level or higher on the backup line.
2. Place a second backup device as high as practical above the knot on the backup line. (Note: if a second backup device is not available, a midline knot may be used to secure the technician temporarily.)
3. Remove the first backup device from below the knot on the backup line.

To pass a knot in an access line while ascending using an ascending system, follow these steps:

1. Ascend to the knot using the normal technique.
2. If necessary/appropriate, pass the knot in the backup line as described above.
3. Perform a changeover from ascending system to descending system, as previously described.
4. Remove the handled ascender with footloop from below the knot, and replace it on the access line above the knot.
5. Ascend on the descender so that the descender is as close as possible to the knot and lock off.
6. Step up in the footloop and connect chest ascender to the working line above the knot.
7. Sit back into the harness, with weight supported by the chest ascender.
8. Remove the descender from the access line.
9. Continue ascending on the ascent system to the desired height.

Descending past a knot is addressed in Chapter 9.

Negotiating an Edge or Obstruction While on Ascent

If the technician is using a rope ascent system to climb past an edge, or to access a rooftop, platform or other working surface, the point where the rope goes over the edge creates an obstacle that must be passed safely. As long as the technician is suspended below the edge, the tension on the rope will keep it compressed against the edge. Moving their ascenders past this edge safely is not complicated, but it must be done properly to avoid compromising safety. Always ensure that the rope is appropriately protected against abrasion or edge damage.

To pass an edge or obstruction on ascent, follow these steps:

Climb up as close as possible to the edge, so that the chest ascender is as high as it will go without impeding the function of the handled ascender.

1. Sit comfortably in the harness, suspended by the chest ascender.
2. Move the backup device past the edge on the backup safety line.
3. Remove the handled ascender from below the edge/obstacle, and replace it on the access rope above the edge. Note: temporary removal of the handled ascender will leave the technician suspended by his chest ascender. This is permissible as long as the chest ascender remains under tension, and there is no fall potential.
4. Step into the footloop and inch the chest ascender past the edge. NOTE: this may take several small moves, depending on the distance and nature of the edge.
5. Continue climbing.

Always protect the ropes from sharp edges and ensure that any necessary rope protection is in place before allowing the rope to make contact with the edge. If edge protection is in place, any necessary manipulation of the edge protection should be performed in a manner that ensures continuous protection of the rope.

If the edge is particularly long or difficult, or if it requires a significant move to get past, it may be necessary to change over to the descending system before moving past it. The deciding factor should be whether the potential exists for a fall or a significant impact force negotiating the edge. In this case, the following procedure may be a more appropriate choice:

1. Ascend to the edge using the normal technique.
2. Move the backup device past the edge on the backup line.
3. Perform a changeover from the ascending system to the descending system.
4. Ascend on the descender so that the descender is as close as possible to the edge without negatively interfering with the performance of the handled ascender.
5. Remove the handled ascender with footloop from below the edge, and replace it on the access line above the edge.
6. Continue ascending until the descender is as close as possible to the edge, then lock off the descender.
7. Step up in footloop and connect the chest ascender to the working line above the edge (and above edge protection, if applicable).
8. Sit back into the harness, with the weight supported primarily by the chest ascender.
9. Remove the descender from the access line.
10. Continue ascending on the ascent system to the desired height.

Passing a Deviation

Deviations, introduced in Chapter 9, involve running the primary access line and the secondary backup line through an intermediate anchor point for the purpose of redirecting the system into a different fall line. Normally, the rope angle is less than 15° from vertical, but this bend in the rope creates an obstruction that the technicians must move past without putting themselves into either a fall or a swing fall potential. Because deviation anchors are often not full strength, the technician should use techniques that avoid committing fully to it as shown in Figure 10-6.

Passing deviations while on descent are discussed in Chapter 9. Passing a deviation is easiest if it is rigged with a sling that is of sufficient length – preferably just a bit longer than a cow's tail. At the end of this sling is a locking carabiner, through which the access and backup ropes are run.

> **NOTE**
>
> If the technician anticipates also having to descend back down the deviation that they are ascending through, he should tie a large, bulky knot into the rope at ground level to facilitate passing the deviation on descent.

To pass a deviation on ascent:

1. Ascend to the deviation anchor.
2. Connect harness waist attachment point directly into the deviation anchor with a short positioning lanyard to maintain position.

FIGURE 10-6
Maintain connection to both primary and secondary systems while passing a deviation anchor. Vertical Rescue Solutions by Pigeon Mountain Industries, Inc.

3. Using a separate, shorter anchor sling, clip the access and backup lines into the deviation anchor with a carabiner, below the technician.
4. Unclip the technician's positioning lanyard.
5. Remove access and backup line from original anchor sling.
6. Use the tail of the rope to control horizontal movement of the technician toward the new fall line.

Keep in mind that the technician must always maintain connection to the primary access system (ascending system) and to the backup safety system while ascending past a deviation anchor (Figure 10-6). The direct connection made between the technician and the deviation anchor is for positioning only and it is generally prudent to not rely upon this as a full point of protection.

Passing a Re-anchor (Rebelay) on Ascent

A rebelay differs from a deviation in that rebelay anchors are always full strength, there are separate rebelay anchors for the access system and the backup system, and the access line and backup line are connected to the rebelay anchors with knots rather than simply being passed through a carabiner. Rigged properly, there will always be some sag in the rope between the primary anchor and the rebelay anchor. The horizontal distance between the primary anchor and the rebelay anchor determines whether the system is considered a long-rebelay or a short-rebelay, and each of these is negotiated slightly differently.

CHANGEOVERS

FIGURE 10-7
In a Short Rebelay, the technician may opt to use only two lines for ascent.

The short rebelay, shown in Figure 10-7, is used primarily for relieving rope weight and reducing the bounce effect on long drops, and for separating multiple technicians on a line. It generally involves less than 15° (or 6 ft) offset between the rebelay anchor and the main anchor. While there may be four sets of vertical ropes in the short rebelay, the technician really needs only to use three sets for the transition. Using all four sets of vertical ropes, as in a long rebelay, is also acceptable – this just involves an extra step, and a little more effort.

As with many other maneuvers, if the technicians are approaching the short rebelay from below while on ascent, they should perform a changeover just below the rebelay anchor if necessary to ensure that they begin the exercise suspended from their descender, just below the rebelay.

To negotiate a short rebelay from below, follow these steps (Illustrated in Figure 10-8):

1. Using a properly rigged system, ascend to the first set of anchors on the lines marked (1) in Figure 10-7.
2. Perform a rope-to-rope transfer from the first set of lines to the far-side of the rebelay loop of the access line (marked (2) in Figure 10-7), so that the technician is suspended from the far-side of the access line's rebelay loop on the ascending system.
3. Remove the first backup device from the initial vertical backup line.
4. The technician should now be suspended from the far side of the two rebelay loops on his/her ascending system with the backup device on the backup line.
5. Continue ascending to the desired height.

The practice of descending back down through this system is addressed in Chapter 9.

Of course, depending on the circumstances, the technician may approach the short rebelay from another perspective; for example, beginning from the top, or from a different side. Or, it may be necessary to move only part way through the short rebelay and then return to the point of origin.

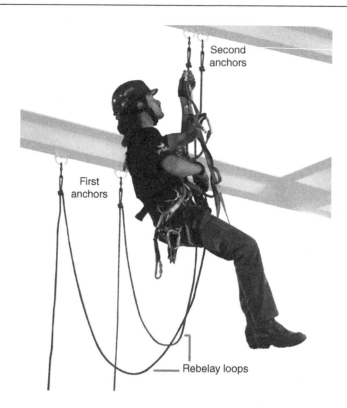

FIGURE 10-8
Technician will use just two sets of lines to negotiate the short rebelay, as shown. Courtesy of Pigeon Mountain Industries, Inc.

The sequence above may be extended, truncated, or otherwise modified to accommodate specific circumstances. The point of the exercise here is that in the case of a short rebelay it is not necessary (although it is acceptable) to use both legs of the interior loop to make the transition from one side to the other.

Not all rebelays are short. A long rebelay is one in which the second set of anchors results in an offset greater than 2 Meters (6 ft) from the main set of anchors, as shown in Figure 10-9.

To negotiate all the way through a long rebelay, the technician must actually use all four sets of ropes, as numbered in Figure 10-10.

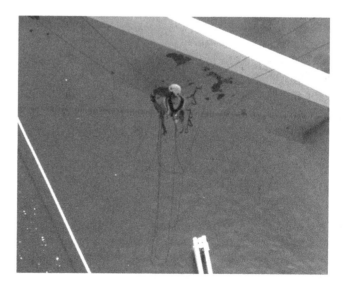

FIGURE 10-9
Long Rebelay involves offset of greater than 6 ft. Courtesy of TranSystems Corporation.

CHANGEOVERS

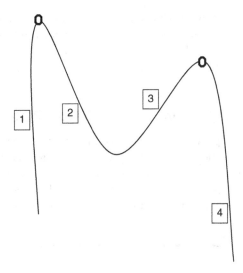

FIGURE 10-10
Long Rebelay.

One method for negotiating the long rebelay is as follows:

1. Using a properly rigged system with a secondary rope access backup, ascend the first set of ropes (marked #1) to the main anchor.
2. Connect a second backup device to the near-side of the rebelay loop of the backup line (marked #2).
3. Perform a rope-to-rope transfer from the vertical access line to the near-side of the rebelay loop of the access line (marked #2), so that the technician is suspended from the near-side of the access line's rebelay loop on the descent system.
4. The technician should now be suspended from the near side of the rebelay in descent mode with backup also attached to the near side of the rebelay.
5. Collect the far side of the access line's rebelay loop (marked #3) and perform another a rope-to-rope transfer, attaching the ascending system to it. Ensure that it facilitates travel in an upward direction.
6. Using the descending system on one set of lines and the ascending system on the other set, maintain consistent tension between the two to avoid swing fall while descending/ascending across the span from set #2 to set #3 of lines.
7. When the technician has descended to directly below the rebelay anchor, he/she will be left suspended in the ascending system (with backup) from the far-side of the rebelay loop (marked #3).
8. Perform another rope-to-rope transfer, this one from the far-side of the rebelay loop to the final set of vertical lines (marked #4).
9. At the end of this series, the technicians will be left suspended on their descender. If further ascent is required, an additional changeover here may be appropriate.
10. Ascend or descend this set of ropes, as applicable to circumstances.

Of course, depending on the circumstances, the technician may not be moving all the way through the system, but may only be using three of the legs, or in some cases just the loop, for access. In such cases, the above sequence may be truncated and modified accordingly. The key here is to ensure that both sides of the interior loop are used for the transition. This differs from the short rebelay, where it is acceptable to use only one side of the interior loop for the transition.

For training purposes, if the technician is capable of moving through all four legs it is quite likely that they will be capable of any likely variation on the system.

Transitioning Off Rope from Ascent, Onto a Platform

If the technicians use a rope ascent system to climb from a low point to a platform or a working surface above, they will need to ensure that safety is maintained until they are safely out of the hazard zone. This may entail using appropriate fall protection methods such as their rope access backup, fall arrest devices, restraint, barriers, or other techniques. Personal ascenders should never be relied upon for fall arrest.

10-5 SUMMARY

Ascending and descending provide the foundational skills for rope access, and every technician should be capable of transitioning effectively back and forth between the two. The systems and methods discussed in this chapter should be practiced under the direction of a competent trainer, and verification of skills should be attained through evaluation and certification before the technician attempts to perform work at height.

Whatever predicament the technicians may get themselves into, they should also be able to get themselves out of it if something goes wrong. Likewise, technicians should always be adequately prepared to rescue a coworker who might become incapacitated. Rescue is further addressed in Chapter 13.

CHAPTER 11

Advanced Techniques

By the end of this chapter, you should expect to understand the fundamental elements regarding the use of:

- Belays
- Aid climbing
- Lead climbing
- Twin lanyard climbing
- Raising and lowering systems
- Cross-haul techniques
- Tensioned ropes
- Powered systems

The previously addressed rope access techniques, including ascending, descending, rope-to-rope transfers, and passing through deviations and other obstructions, comprise the vast majority of techniques used most often by rope access technicians. However, the key to safety in rope access lies in the ability to maintain protection under circumstances that might otherwise be deemed "infeasible" to protect.

Indeed, in the world of rope access, maintaining a connection to both a primary and a backup system at any point in time is a nonnegotiable requirement.

To maintain this constant protection, it becomes necessary to explore rigging and access methods that are a little less conventional, and to learn to apply these methods in a manner that is both safe and practical.

As with all techniques, these methods should be used only after a risk assessment has been performed by a competent person, and appropriate planning and equipment choices have been made. Worker safety and provision for rescue in the event of a fall, are paramount.

Professional Rope Access: A Guide To Working Safely at Height, First Edition. Loui McCurley.
© 2016 John Wiley & Sons, Inc. Published 2016 by John Wiley & Sons, Inc.

11-1 BELAYS

The term belay means, simply, to stop or arrest. The purpose of a belay in the context of a rope access system is to catch the load in case of a failure in the primary system. In effect, the idea behind every rope access backup system is to provide a "belay" to the technician in case of primary system failure. Of course, we normally do not call the backup system a belay, we call it a rope access backup system.

At times, however, use of a conventional rope access backup may not be feasible for whatever reason. In these situations, suitable alternatives may include either **passive belay**, or **active belay** systems. When using such nonconventional backups, we do typically refer to these as a belay system.

A passive belay is one that activates automatically, such as a rope grab, while an active belay usually consists of a secondary braking device operated by another technician. When using either type of belay, the technician should also have a primary system in play; whether it is a suspended rope, a walking/working surface, a ladder, or some other means of support. The belay is intended only to provide a secondary level of protection – that is, one-hundred percent fall protection.

Use of a passive belay is arguably the easier approach because it moves with the load automatically, and locks off in the event of sudden movement. One example of a passive belay is a self-retracting lanyard (SRL) shown in Figure 11-1.

FIGURE 11-1
Self-retracting lanyard (SRL). Courtesy of Reliance Industries, LLC.

BELAYS

An SRL is typically contained within a closed box, with a safety line wound around a tensioned block inside. A centripetal brake causes the block to lock in the event of a sudden increase in velocity. For this reason, care must be taken in using an SRL to not move too quickly, or the device will lock off. Once locked, tension must be removed from the device to release the brake.

The active belay, shown in Figure 11-2, requires a little more training and experience for safe use, but is arguably more versatile. In an active belay, the technician (or load) is connected directly to a rope, the running end of which is passed through a braking device that is anchored to a solid point. Selection of an appropriate belay device is addressed in Chapter 4. The belay device is actively managed by a second technician, called a "belayer." As the load moves, the belayer feeds the belay rope out while maintaining light tension on it. If at any time the primary system supporting the load fails, the belayer would increase tension on the belay rope to prevent it moving through the belay device, thereby catching the fall.

A belay system may be used to protect a technician who is moving through potential fall hazards, such as lead climbing or when suspended by rope. A belay system may also be used to protect multiple technicians, such as during a rescue.

Proper belay technique requires good rope management skills on the part of the belayer, as well as good communication between the belayer and the person being belayed. Where possible, the belay rope should be stacked or flaked prior to

FIGURE 11-2
Active belay system.

FIGURE 11-3
Belayer position relative to device. Courtesy of ClimbTech.

commencing movement. This helps to ensure that the rope feeds easily and cleanly through the belay device during use, without the belayer being distracted by twists and tangling. The belay device should be connected to a secure anchorage rather than directly to the belayer, so that the belayer can easily remove himself from the system in the event of an emergency. The belayer should position himself relative to the belay device in a manner that facilitates proper operation of the device. Position may vary depending on which device is being used (Figure 11-3).

While an active belay arguably requires greater training and skill than a passive belay for effective operation, it does offer more flexibility in use, particularly in complicated environments. An active belay also permits a modicum of control by the belayer, and greater maneuverability for the person being belayed. Even with an autolocking braking device, an active belay relies on human performance to function effectively. For this reason, technicians must be adequately trained in belaying prior to using this method.

When using a belay system, the person being belayed should always be the focus of all commands and prompts. While the belayer rigs the belay device to an appropriate anchorage and feeds the rope properly through it, the person who will be belayed ties directly to the rope. Before exposing themselves to any potential fall hazard, belayed persons should instigate communication with the belayer to ensure that they are secure. An appropriate series of commands might be something along the lines of:

Belayed Technician: On Belay?
Belayer: Belay is On

Belayed Technician: Climbing
Belayer: Climb away
Belayed Technician: Stopping
Belayer: Stop
Belayed Technician: Off Belay
Belayer: Belay is Off

A belay is appropriate for inclusion wherever:

- the technician is using some other means for primary support,
- there is a high likelihood and/or consequence of failure of the primary system, and
- the protection provided by the belay outweighs the potential hazards that the belay might create.

When using moving ropes to raise, lower, or move a load horizontally, always consider whether to incorporate an active belay into the system.

11-2 AID CLIMBING

Aid climbing is a method used by rope access technicians to access an expanse where ropes have not yet been set, and where the structure itself is not conducive to climbing. Aid climbing allows the technician to move in any direction on a structure by using the structure itself to support a series of adequate anchorages, which may be placed during upward progress as necessary. Lanyards are used to maintain continuous connection of at least two independent systems at all times, one for suspension and the other for backup. During transitions, there are at times three points of connection in place.

Since a rope access technician is typically equipped with three cow's tails at all times, it is prudent to learn to use these to perform the aid maneuvers. In preparation for the aid climb, the technicians should set up their equipment with an etrier (or footloop) on the end of each long cow's tail, and a carabiner on the end of their short cow's tail. The long cow's tails are typically attached to the technicians' sternal D ring, while the short cow's tail is attached to their waist D-ring. If the waist-level cow's tail is longer than what the technician might like, the chest ascender may be used to "choke up" on it as desired. Rigging for a horizontal aid climb is shown in Figure 11-4.

Prior to engaging in aid maneuvers, the route should be planned to ensure adequate protection, and lanyard lengths should be optimized for ergonomic efficiency.

To use aid climbing techniques to move horizontally across a structure is known as an "aid traverse". To perform an aid traverse, the technicians should:

1. Attach their short cow's tail to the first aid anchor
2. Attach one long cow's tail (with footloop) to the same (first) aid anchor
3. Attach the second long cow's tail (with footloop) to the second aid anchor
4. Stand with one foot in each footloop, and unclip the short cow's tail from the first aid anchor and clip it to the second aid anchor
5. Sit down into the harness, so that they are suspended by the short cow's tail from the second aid anchor

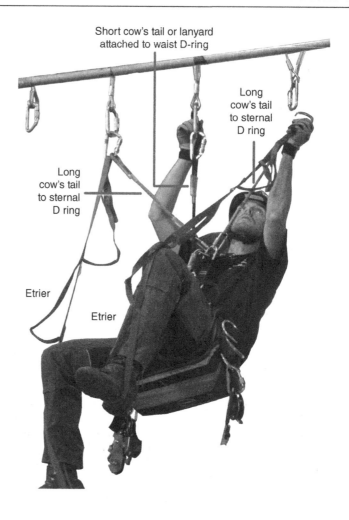

FIGURE 11-4
Technician aid climbing horizontally. Courtesy of Pigeon Mountain Industries, Inc.

6. Move the long cow's tail from the second aid anchor to the third
7. Move the long cow's tail from the first aid anchor to the second
8. Stand with one foot in each footloop, and transition the short cow's tail from the second aid anchor to the third.
9. Continue to progress across the structure in this manner, always maintaining at least two separately anchored points of connection to the structure at all times.

When the technicians reach the end of the traverse and/or has completed their work there, they may egress either by making their way back across the traverse, or by transitioning to another system. In some cases, this may be an appropriate place to use a pull-through system for descent, as described in Chapter 8.

Aid climbing may also be used in a similar manner for climbing vertically up or down a structure, as shown in Figure 11-5.

When moving vertically, fall potential should be monitored carefully. Special care should be taken to ensure that any fall potential is mitigated to less than 2 ft and under 900 lb potential force. If the fall potential cannot be thus mitigated, a force-absorbing lanyard should be used. However, in this case the additional clearance distance required must considered.

Anchorages for lead climbing, whether vertically or horizontally, should be selected carefully to ensure that they are of sufficient strength and that they are

LEAD CLIMBING

FIGURE 11-5
Aid climbing vertically on a structure, trailing ropes for later use. Courtesy of Trask Bradbury.

not too far apart. The technician is always connected to at least two anchorages at all times, one for suspension and the other as backup. Any anchorage point(s), configuration(s) and material(s) may be used for the anchorage, as long as it is of sufficient strength to meet local regulatory requirements and to perform the task at hand with adequate safety margin. An example of a series of anchorages used for lead climbing is shown in Figure 11-6.

11-3 LEAD CLIMBING

Lead climbing is a technique, used only in rare circumstances by specially trained technicians, which allows the technician (the "climber") to directly climb a structure while being protected by an appropriate harness and safety line, belayed by a second technician (the "belayer"). An example of a lead-climbing arrangement is shown in Figure 11-7.

The technicians climbing the structure may move in any direction along the structure, so long as they are adequately protected against significant impact and/or

FIGURE 11-6
Anchorages, whether for aid climbing or for lead climbing, must be adequate to the task. Courtesy of Trask Bradbury.

swing fall. This requires that the climber place intermediate anchorage points along the climb path as they proceed, and that they clip the safety line into these anchorage points before climbing above them. As the climber moves upward, they establish these intermediate anchor points at a frequency that minimizes the extent and severity of a fall and passes the rope through them before continuing on. In this manner they may ensure that the maximum potential fall distance is no more than 2× the distance from their position to the nearest anchorage point. The technician can minimize potential fall distance by keeping the anchorage points closer together. The desire for a short fall distance must be weighed against the amount of energy and effort required to place each anchorage point.

During the lead climb, the safety of the climber is in the hands of the belayer whose responsibility it is to feed the rope out so that there is minimal drag on the climber. In the event that the climber falls the belayer must grasp the rope firmly to activate the belay device and stop the rope from moving. For this, adequate belay device and not bare hands must be used. An example of an auto-locking belay device is shown in Figure 11-8.

Clearly, this scenario raises the potential for fall distances to exceed 2 ft, which in turn may increase potential impact forces on the climber in the event of a fall. To mitigate the hazard, an appropriate dynamic climbing rope should be used as the protective line. To ensure use of a rope that is specifically designed for this purpose,

LEAD CLIMBING 205

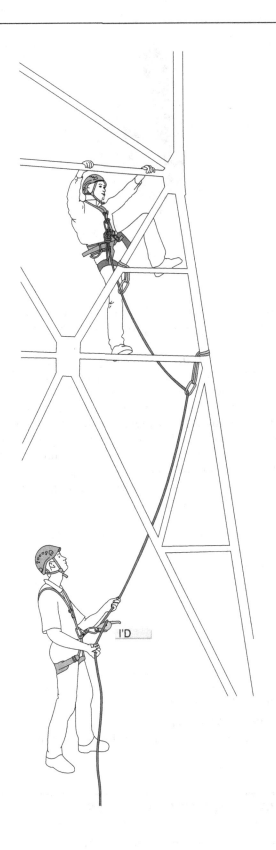

FIGURE 11-7
Lead climbing. Courtesy of Petzl.

FIGURE 11-8
Autolocking devices work well for belay. ©BEAL.

a climbing rope that is designated for life safety purposes and that is certified as meeting an adequate dynamic climbing rope standard such as UIAA 102 must be selected.

Alternatively, a force-absorbing lanyard may be used to mitigate forces. Again, however, it should be borne in mind that the deployment of a force-absorbing lanyard will necessitate additional clearance distances.

Because lead climbing does not meet the strict definition of rope access (two-rope system), it is not really considered to be a rope access technique. In this case, the structure itself serves as the "primary access system" while the dynamic belay serves as the "backup system." Although lead climbing is not strictly rope access, it is a viable and useful skill when used by experienced technicians, and can be indispensable in maintaining protection where resources are limited.

Before using this method for access:

1. Plan the route carefully to ensure adequate protection will be available along the way, at optimum spacing.
2. Ensure adequate clearances.
3. Preplan for rescue in case of a fall.
4. Mitigate exposure of the line to sharp edges.
5. Ensure that both the climber and the belayer are adequately trained and equipped.

During the process, the belayer should take care to:

1. Minimize slack in the safety line, without "tugging" on the climber
2. Operate the belay device properly
3. Pay close attention to the needs of the climber, and respond accordingly.

11-4 CLIMBING WITH TWIN LANYARDS

Another indispensable technique that is not classified strictly as rope access is that of climbing with twin lanyards. This technique may be used to progress along a structure where adequate anchorages are available to alternately clip and remove

FIGURE 11-9
Climber with twin lanyards. Courtesy of Petzl.

lanyards while moving forward, so that if the technicians were to slip and fall they would neither impact the ground nor experience significant impact force.

Twin lanyards are usually used attached to the dorsal attachment point of an appropriate full body harness, in accordance with local regulatory requirements. A sternal attachment may be used if the lanyards are short enough, and if local regulation permits. An example of twin lanyard use is shown in Figure 11-9. When permissible, there is arguably a significant advantage to choosing a sternal attachment, as this improves the prospect of the technician being able to regain footing in the event of a fall, thereby facilitating self-rescue. In addition, unlike the dorsal point, use of the sternal attachment reduces the risk of injuring one's face against the structure during a fall.

When climbing with twin lanyards, the structure serves as the primary access system while the lanyards serve as backup. As the climber advances, one lanyard at a time should be attached to the structure. The climber should progress so that the harness attachment is at the same level as where the lanyard is anchored, then reach up and anchor the other lanyard as high as practicable above before removing the first lanyard, and so on.

11-5 RAISING AND LOWERING SYSTEMS

While most rope access systems are designed to be stationary, so that the technician progresses along a fixed rope, it is also possible to use moving ropes to position a technician or other load at a designated location. These systems are useful for positioning a technician who needs both hands free to work, positioning a worker who is not rope access certified, for rescue or for other purposes.

Broadly, there are two aspects to moving a load: raising and lowering. One should not be undertaken without also having the capability to perform the other. This will help prevent getting a load into a predicament from which it cannot be readily removed. These operations are useful in a wide variety of circumstances, from low angle to steep, up to and including vertical and overhanging situations. Friction (discussed in Chapter 7) is the primary means used to control the speed and direction of the load during descent, and mechanical advantage (also discussed in Chapter 7) used for raising.

There are many effective and appropriate means by which to execute these tasks, so this text will seek to focus on the fundamental concepts. Given this approach, the technician should be able to approach each situation with fresh perspective, and then apply the resources that are available to achieve the desired goal. While the means and methods of applying these concepts are virtually unlimited, there must always be the caveat that forces and fall potential must be mitigated, and that the load should always be protected from falling in any direction by at least two lines – a primary access line and a backup.

11-6 SYSTEMS FOR LOWERING

Working from above to lower a load into place is an approach that uses gravity to the fullest, thereby reducing the amount of exertion required. Generally, at least three technicians are required: one to operate the system (called the "brakeman"), one to operate the belay (called the "belay") and one to be lowered (called the "load"). During the operation, the load should be in command, and should direct the brakeman to proceed in their direction. If the load is an inanimate object, a technician who is in position to be able to see the load should be in command.

In terms of equipment, a lowering system requires at least an anchorage, rope, braking device, and an adequate selection of connectors. The anchorage for a lowering system should be of sufficient strength to withstand the desired load (with adequate safety factor), and should be in compliance with local regulation. It should be positioned above and in line with the desired point of access, but without creating a hazard, and a braking device should be connected to the anchorage.

To lower a load, follow these steps:

1. Attach the working end of the rope to the load.
2. Attach a belay, vertical lifeline, or other secondary backup system to the load.
3. Brakeman, belayer, and load/command confirm communications terminology, and confirm that all are ready.
4. Brakeman and belayer pull ropes back through their respective braking devices to remove any slack, creating tension between the anchorage and the load.
5. Using a command such as "down slow," load directs brakeman to begin to lower the access line slowly through the brake.
6. Brakeman complies. Belayer follows suit.

Perhaps the most difficult part of a lowering operation, or in fact any operation that involves multiple technicians, is ensuring effective communication. When a technician is on a rope, the technician's communication must take precedence over all others – especially when someone else is controlling the descent. Principal communications should be between the brakeman and the load/command, either by direct voice or by radio.

It is best to use simple, predetermined commands during lowering or other moving-rope operations. Agreeing in advance to a specific set of words to be used is helpful. The words selected should communicate the appropriate information *while altering cadence, intonation, and sound* enough to prevent misunderstanding. Ideally, communications should follow a statement-and-response pattern.

One example of a good set of commands might be as outlined here:

Brakeman: Ready for Lower on Your Command! (9 syllables; last sound "...and!"; statement)
Technician: Am I on belay? (5 syllables, last sound "...lay?"; question)
Brakeman: Your belay is on! (5 syllables, last sound "...on!"; statement)
Technician: Down slow! (2 syllables, last sound "...oh!"; statement)
Brakeman: Lowering (2 syllables, last sound "...ing!"; statement)
Either Person: STOP – STOP! (2 syllables, last sound "...op!"; statement)
Either Person: Stop. Stop. Why? (3 syllables, last sound "...why?"; question)
Either Person: Preparing to... (Varies)
Technician: I am off belay. (5 syllables, last sound "...lay!"; statement)
Brakeman: Belay is off. (4 syllables, last sound "...off!"; statement)

These commands may be used in any order, as dictated by need/situation. A similar set of commands can be used during a Raising operation:

Brakeman: Ready for Raise on Your Command! (8 syllables; last sound "...and!"; statement)
Technician: Am I on belay? (5 syllables, last sound "...lay?"; question)
Brakeman: Your belay is on! (5 syllables, last sound "...on!"; statement)
Technician: Up slow! (2 syllables, last sound "...oh!"; statement)
Brakeman: Raising (2 syllables, last sound "...ing!"; statement)
Either Person: STOP – STOP! (2 syllables, last sound "...op!"; statement)
Either Person: Stopping. Why? (3 syllables, last sound "...why?"; question)
Either Person: Preparing to... (Varies)
Technician: I am off belay. (5 syllables, last sound "...lay!"; statement)
Brakeman: Belay is off. (4 syllables, last sound "...off!"; statement)

Note the variation in the number of syllables and final sounds in each of the series above.

11-7 SYSTEMS FOR RAISING

Whenever a lower is performed, technicians should also be prepared to raise in the event that it becomes necessary. Mechanical advantage, discussed in Chapter 7, can be very useful when it comes to raising a load. The mechanical advantage may be built directly into the access line, or a separate system may be "piggybacked" onto it.

FIGURE 11-10
Lower converted to a direct raise by addition of pulley. Descending device, rigged high in this image, is also used as progress capture. Courtesy of Vertical Rescue Solutions by Pigeon Mountain Industries, Inc.

Some descenders are built with the idea of being able to raise and lower, as needed, and may even capture progress during the raise. Conversion of such a system from lower to raise simply requires placing an appropriate pulley device on the rope downstream from the descender and reeving the rope through it, as shown in Figure 11-10. From here raising a load is simply a matter of pulling.

This simple action creates what is called an in-line 3:1 haul system. Although using a descender as a pulley/progress capture device does generally sacrifice some level of efficiency, system simplification and ease of operation often outweigh the efficiency loss. In this type of system, converting from a lower to raise and back again can be accomplished in a matter of seconds.

Another method of raising a line is to "piggyback" a haul system onto the primary line. Piggybacking refers to the idea of attaching a prerigged haul system to a sturdy anchor at the top, and connecting it onto the lifeline with a substantial rope grab. Pulling on the piggybacked haul system, in turn, pulls on the lifeline, which raises the load (Figure 11-11).

The haul systems described above may be built out to incorporate any numerical value of mechanical advantage that is desired.

Technicians should be trained to be capable of employing haul systems effectively from a platform or ground level, as well as while being suspended from a rope.

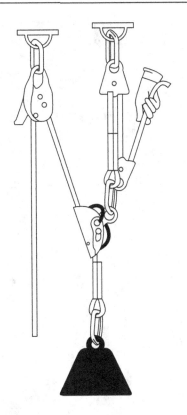

FIGURE 11-11
Piggybacked haul system. Courtesy of Margaret DeLuca.

Because each of these situations involves different challenges, technicians should practice and be familiar with any and all situations in which they might find themselves.

11-8 CROSS-HAUL

Haul systems may also be used to move a load in any direction while in midair simply by using multiple systems simultaneously. The term used to describe this type of system is "Cross-Haul," shown in Figure 11-12.

Especially useful in the event of a complex rescue scenario, cross-hauling may be employed to maneuver a load up, over, under, or through a great variety of obstacles using two or more haul systems rigged to pull from varying directions. The load is effectively moved by alternately applying and releasing tension in each of the systems.

The greatest challenges in a cross-haul scenario involve:

1. Managing cumulative vectored forces
2. Effective communication between technicians operating each of the systems
3. Ensuring adequate backup so that the failure of any one point cannot result in catastrophic failure.

FIGURE 11-12
Cross-haul. Vertical Rescue Solutions by Pigeon Mountain Industries, Inc.

11-9 TENSIONED ROPES

In some cases, rope access technicians may find the use of tensioned anchor lines, in conjunction with raising/lowering systems, to be advantageous for moving people, tools, and materials in both a horizontal and vertical plane, as illustrated in Figure 11-13

There are two general approaches to this. The first is known as a guideline. A guideline is a tensioned rope (called a track line) rigged in a sloping configuration between two anchors. A load may be attached to the track line by means of a pulley, and then raised and lowered along the sloped path using the raising and lowering methods discussed earlier in this chapter.

When the track line is horizontal or near horizontal, it is more appropriately known as a highline.

The more tightly the track line is tensioned between the anchors, the easier the load is to move. Static ropes (described in Chapter 5) provide more consistent tensioning behavior than low stretch. However, the tighter a rope is tensioned, the higher the forces will be at the anchors when a load is applied and travels across the span. Either one rope or two may be used for the track line. For a two-rope track line, a special wide-sheaved pulley, sometimes called a Kootenay pulley, may be useful.

FIGURE 11-13
Tensioned rope systems are useful for overcoming obstacles. Courtesy of Pigeon Mountain Industries, Inc.

In a two-rope track line, both lines should be tensioned and maintained as equally as possible between the two anchors. One anchor is rigged as a fixed anchor while the other serves as a tensioning anchor. If possible, using a high-strength tie-off at the fixed-anchor end will help preserve maximum rope strength and more stable rigging.

At the tensioning anchor, the track line(s) feeds into a mechanical advantage system that is used to tension the track lines. The amount of tension appropriate for the track line depends on a number of factors including the slope of the highline, the anticipated load, and the strength of the system. As a rule of thumb, when the load is at midspan there should be approximately 10% sag in the track lines.

To facilitate movement of the load along the track line, working lines with lowering systems and/or mechanical advantage systems may be attached to the load and operated as needed from each anchor. These working lines may also be used for backup safety, to protect against catastrophic failure in the event that the main line should fail, but in such cases they must be managed effectively to prevent slack.

Great care must be taken to evaluate and constantly manage the angles of rigging, forces at the anchor(s), terminations, and at points throughout the system so as to not exceed the safe working load of equipment. Technicians must have a thorough understanding of forces and rigging principles, such as those discussed in Chapters 7 and 8, before attempting to rig or use a highline. These concepts are further complicated by the fact that as a suspended load moves across a tensioned rope system, the direction and magnitude of the force will change throughout the system.

It is easy to over-tension a highline. The more tightly the track line is tensioned, the more easily the load will be moved – but the greater the forces will be. Forces in an over-tensioned system can be catastrophic. For this reason, the system should be used with the minimum practicable rigging angles, and forces should be calculated and monitored by a competent person throughout the operation of such a system to ensure that the system is not overtaxed.

Of course, consideration must be given to establishing adequate safety backup measures. As with any work-at-height system, the failure of any one component should not result in catastrophic failure.

FIGURE 11-14
Two or more systems may be used simultaneously to facilitate positioning and avoid swing fall potential. Courtesy of Abseilon USA, www.abseilon.com.

11-10 MULTIPLE SIMULTANEOUS SYSTEMS

It is not uncommon for the rope access technician to apply two or more systems simultaneously, as in the use of a haul system along with a guideline, or using two sets of horizontal or near-horizontal ropes simultaneously for positioning as shown in Figure 11-14.

A good understanding of rigging principles is of utmost importance when using two systems simultaneously, as the effects of one system is likely to have at least some bearing on the other. In all cases, the rope access technician will employ a primary system for access and a secondary system for backup. When an additional primary system is added for access, consideration must also be given to adding an additional secondary system for backup safety.

11-11 POWERED ASSIST SYSTEMS

Experienced rope access technicians may find the use of powered assist systems to be beneficial. Sometimes referred to as "powered ascenders," these devices can contribute to greater efficiency and reduced fatigue when performing many of the aforementioned maneuvers, and are especially useful where a great deal of movement is required along the rope path.

Powered assist systems may run on gasoline, AC, or DC power, and may or may not have a built-in comfort-seat. With a powered assist system, technicians can ascend farther and work longer due to the increased comfort and support as well as the lower energy required (Figure 11-15).

Rope access technicians should use only powered assist systems that are designed and approved by the manufacturer to support human loads. Any system used should adequately accommodate ascent as well as descent, and must incorporate fail-safe mechanisms to prevent loss of control.

SUMMARY

FIGURE 11-15
Powered assist system. Courtesy of Sean Cogan.

Only properly trained and certified rope access technicians should use powered assist systems, and they should be used in a manner that allows the technician to abandon the hardware and revert to manual rope access techniques if necessary. Powered assist systems should also be rated to support a two-person load in case of rescue.

Powered assist systems are discussed in greater detail in Chapter 13.

11-12 SUMMARY

While the basic maneuvers of ascending and descending will get a technician safely through the vast majority of rope access jobs, being able to understand and perform advanced techniques is what really sets rope access technicians apart.

The premise of one-hundred percent protection is fundamental to safe and effective rope access. While some may perceive these advanced techniques as pushing the limits of conventional fall protection, it is this very thing – the ability to analyze and effectively rig to protect and even rescue from the most challenging of circumstances – that keeps the rope access technician safe where conventional methods might write a situation off as "infeasible to protect."

As with all chapters in this book, the guidance provided in this chapter is not a replacement for training and hands-on practice under observation. The knowledge gleaned here or elsewhere must be augmented with competent training and hands-on exercise, followed by independent evaluation and affirmation of skills.

CHAPTER 12

Use of Powered Rope Access Devices

- Overview
- Precautions
- System configuration
- Methods – sit-on-top
- Methods – suspension beneath
- Methods – fixed position
- Additional considerations
- Care and maintenance

The use of powered devices in work at height is a relatively new development, but their safe and effective implementation has revolutionized both convenience and efficiency. Especially useful in circumstances where long ascents are necessary, such as wind turbines, bridge inspection, and building maintenance operations, powered devices offer improved safety and efficiency. While the most obvious advantage of powered devices is a marked reduction in effort required for ascending, other capabilities include descending, use as a comfort seat in a fixed position, and operation as a fixed winch. Using powered devices, as shown in Figure 12-1, the worker is able to ascend more quickly with less exertion, stay in position longer, and work in relative comfort.

12-1 PRECAUTIONS

Powered devices are not a substitute for appropriate training and experience in rope access. In fact, these devices should be used only by competent, certified rope access technicians who also possess additional training relative to the powered equipment they use.

Professional Rope Access: A Guide To Working Safely at Height, First Edition. Loui McCurley.
© 2016 John Wiley & Sons, Inc. Published 2016 by John Wiley & Sons, Inc.

FIGURE 12-1
Use of powered devices can increase efficiency of rope access workers. Courtesy of Sean Cogan.

Powered devices must be properly cared for and should always be inspected by a competent user prior to use. Maintenance requirements may vary depending on the make, model, and manufacturer recommendations, so the product instructions should always be read prior to use. This equipment should never be altered or modified without written consent from the manufacturer. Because regulatory requirements may vary between jurisdictions, a little research on health and safety rules in the country/locality where the device is to be used is generally a good precaution.

Like any powered tool, the mechanical performance of a powered device is directly related to the power source. Some devices may be gasoline operated, while others may run on electricity or battery. Proper engine maintenance guidelines should be followed in accordance with manufacturer's instructions, and appropriate ventilation and adherence to atmospheric precautions should be maintained during use.

Prior to using any powered device for rope access or other work at height, the employer should be clear as to how the equipment is to be used, and the use and limitations of the powered equipment must be incorporated into the Comprehensive Managed Fall Protection Plan. Specific information regarding use of this equipment should be included in both the Rope Access Plan and the Rescue Plan.

Whenever using a powered rope access device, the rope access technician must wear appropriate PPE suitable to the environment and task at hand, as discussed in Chapter 5 of this book.

12-2 CONFIGURING THE DEVICE INTO THE SYSTEM

When using a powered system within the context of typical rope access, the powered device is connected to the primary progress line, and the technician is attached to the device in accordance with the manufacturer's instructions.

As with any use of rope access, the technician should be properly rigged to both the progress system and an appropriate backup system, as shown in Figure 12-2. Any rope access backup device is fine, as long as it is one that will withstand foreseeable potential forces in the event of a primary line failure. The technician should be connected to the backup system in the usual manner, that is with a cow's tail and rope access backup device, usually to the sternal attachment of the harness.

In this configuration, the technician can use the power of the engine to ascend the rope and can simply use friction to descend.

A typical powered device will include at least the following features, as shown in Figure 12-3:

Chassis – The frame in which the device is housed

Power Source – Generally either electrically powered or battery powered or gasoline powered engine

Starter Mechanism – A pull start or electric switch

Motor – To provide energy to turn the capstan

Capstan – A drum around which the rope is reeved several times to achieve friction

Self Tailing Jaws – Mechanism used to secure the rope in place after its final wrap around the capstan

Control Switch – A button, lever, or other mechanism that control the transmission to start/stop the capstan from moving

Fail Safe Mechanism – A kill switch that disengages the control switch in the event that the operator lets go

Descent Release – Lever or other mechanism that disengages self-tailing jaws and permits the rope to slide around the capstan for descent

FIGURE 12-2
When a powered device is used, the powered device typically serves as the technicians connection to the primary system. As with any rope access system, a backup must also be used. Courtesy of Sean Cogan.

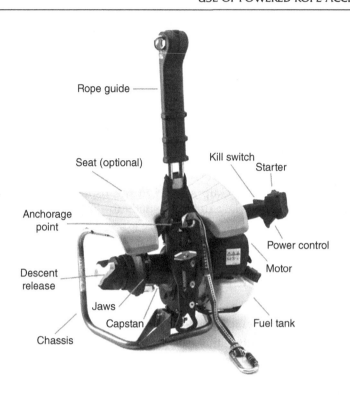

FIGURE 12-3
Common features of powered devices for rope access. Courtesy of Sean Cogan.

Anchorage Point(s) – Approved load-bearing connections where the operator(s) may clip in and/or where device may be anchored

Seat – Some units offer an optional seat on which the operator may sit when using the device.

While specific details regarding the rigging of the rope onto the device may vary depending on the device used, the function of most powered rope access devices requires that the rope be wrapped several times around the capstan and secured in the manner specific to that device. For sit-on-top devices, it may also be necessary to pass the rope through a rope-path guide as shown in Figure 12-4 to ensure that it tracks properly onto the capstan drum.

With the rope wrapped properly around the capstan, the capstan turns and pulls the unit upward along the rope when the unit is powered and the capstan is engaged. For descent, the power to the capstan is suspended and the technician operates a release mechanism to permit the rope to slide around the capstan in a controlled manner.

The optimum number of wraps that the rope should take around the capstan will depend on the design and specifications of the device, how much load will be placed on it, and the condition of the rope and contact surfaces. Manufacturer's instructions will provide details that are specific to the device, but it is generally good practice to perform a small "test" of the device after installing it on the rope to ensure that it is working smoothly. To do this, simply prepare for ascent as described in the product instructions, ascend up to a height of 1 m or less, and then descend in accordance with product instructions. If necessary, the number of wraps may be adjusted and the test repeated before commencing work.

While specific procedures for operation may vary between devices, this text will provide a general overview on two common configurations in which powered devices may be used.

FIGURE 12-4
For sit-on-top devices, the rope many need to pass through a guide. Courtesy of Sean Cogan.

12-3 CONFIGURATION 1 (SIT ON TOP)

The sit-on-top method (Figure 12-5) of using a powered rope access device is especially useful when the technician anticipates to remain on the rope for an extended period of time. Some body types will find this approach to be more comfortable and stable, while other body types may prefer method 2, that is suspension beneath the device, as shown later in this text.

To ascend using a powered device with a sit-on-top feature, follow these steps:

1. Rig the rope access system with two ropes, a primary and a secondary, as discussed in Chapter 8. Each system should have its own separate anchorage(s).
2. Connect the Powered Rope Access Device onto the working line in accordance with product instructions.
3. Sit comfortably on the device seat and connect to the powered rope access device as directed in the product instructions.
4. Connect the technician via sternal attachment to the secondary (or backup) rope in accordance with appropriate rope access techniques.
5. Activate power to the device, as directed in the product instructions.

FIGURE 12-5
Worker using sit-on-top method. Courtesy of Sean Cogan.

6. Engage power to the capstan so that it turns and pulls the unit (and the technician) up the rope.
7. It may be necessary to "tail" the rope – that is, to provide slight back-pressure to the loose rope beneath the device – during ascent.
8. Monitor backup device during ascent to minimize potential fall distance.

To descend using a powered device with a sit-on-top feature, follow these steps:

1. Rig the rope access system with two ropes, a primary and a secondary, as discussed in Chapter 8. Each system should have it's own separate anchorage(s).
2. Connect the Powered Rope Access Device onto the working rope in accordance with product instructions.
3. Sit comfortably on the device seat and connect to the Powered Rope Access Device as directed in the product instructions.
4. Connect the technician via sternal attachment to the secondary (or backup) rope in accordance with appropriate rope access techniques.
5. When the task is completed and descent is desired, power to the device may be deactivated as directed in the product instructions.
6. Grasp the free end of the rope and apply hand-tension.
7. Release the rope grab mechanism to allow the rope to travel around the capstan.
8. Continue to grip the free end of the rope by hand, releasing slowly to apply consistent rope tension around capstan and control descent speed.
9. Monitor backup device during descent to prevent unintentional lockup and to minimize potential fall distance.

FIGURE 12-6
Worker suspended beneath powered rope access device. Courtesy of Sean Cogan.

12-4 CONFIGURATION (SUSPEND BENEATH)

To ascend while suspended beneath a device, as shown in Figure 12-6, follow these steps:

1. Rig the rope access system with two ropes, a primary and a secondary, as discussed in Chapter 8. Each system should have its own separate anchorage(s).
2. Connect the powered rope access device onto the working rope in accordance with product instructions.
3. Connect front waist harness attachment to the device in the manner prescribed by the product instructions, with the device at approximately

shoulder level, so that the technician is suspended comfortably beneath, but within easy reach of, the device.
4. Fasten the technician to the secondary (or backup) rope in accordance with appropriate rope access techniques.
5. Activate power to the device, as directed in the product instructions.
6. Engage power to the capstan so that it turns and ascends up the rope.
7. Grip the free end of the rope by hand to apply consistent tension around the capstan during ascent.
8. Monitor backup device during ascent to minimize potential fall distance.

To descend while suspended beneath a device, complete these steps:

1. Rig the rope access system with two ropes, a primary and a secondary, as discussed in Chapter 8. Each system should have its own separate anchorage(s).
2. Connect the powered rope access device onto the working rope in accordance with product instructions.
3. Connect the front waist harness attachment to the device in the manner prescribed by the product instructions, with the device at approximately shoulder level, so that the technician is suspended comfortably beneath, but within easy reach of, the device.
4. Attach the sternal D ring of the technician's harness to the secondary (or backup) rope in accordance with appropriate rope access techniques.
5. When work is completed and descent is desired, power to the device may be de-activated, as directed in the product instructions.
6. Grasp the free end of the rope and apply hand-tension.
7. Release the rope grab mechanism to allow the rope to travel around the capstan.
8. Continue to grip the free end of the rope by hand, releasing slowly to apply consistent rope tension around the capstan and to control descent speed.
9. Monitor backup device during descent to prevent unintentional lockup and to minimize potential fall distance.

12-5 USING THE DEVICE FROM A FIXED POSITION

Most powered rope access devices may also be used from a fixed position to raise and lower a load (such as a person), as shown in Figure 12-7.

Use of the device in this manner requires at least two operators: one (the "ground operator") to remain at the fixed location with the device to operate the raise and lower function, while the other (the "suspended technician") is suspended at the working end of the rope. The person suspended is, in effect, the "load" and it is this individual who generally performs the work. In a rescue scenario, there may be two persons suspended at this end of the rope.

Close communication is required during such operations, and, if possible, a clear line of sight should be maintained between the operator and the suspended load. The person suspended is always the one in command, and for safety reasons the device operator should take directions from this person.

A baseline of understanding for effective communications should be established to address any risk exposure. The pattern of communication may be similar to that used for the raising/lowering operations discussed in Chapter 11. In fact, the overall

FIGURE 12-7
Using the device from a fixed position. Courtesy of Sean Cogan.

process is quite similar to that which is discussed in the raising and lowering section in Chapter 11, the main difference being that it is a powered device, rather than a manual raising/lowering system that is used to move the load.

Prior to using the powered rope access device from a fixed position, a secure anchorage should be selected that has sufficient working space for the device to rest securely on the ground without being crowded or be in the way of the personnel. Note that the location and position of the device may change when it is placed under the load, so this has to be analyzed in advance and prepared for. The accelerator control should be within easy reach of the ground operator, and the rope path should be unobstructed and free from hazards. It should be ensured that the rope leaves the device in appropriate alignment and at an appropriate angle, in accordance with manufacturer recommendations.

To move a suspended technician(s) or other load using a powered rope access device, proceed as follows:

1. Install the standing end of the primary rope on the device capstan in accordance with manufacturer instructions.
2. Connect the device securely to a suitable anchorage.
3. Ensure that the device has sufficient power available.
4. Connect the technician who is to be suspended to the working end of the rope.
5. Connect the technician who is to be suspended to an adequate backup safety system.
6. Technicians use established communication methods to verify that the rope and system are engaged and ready (i.e., *Ready for operation on your command; On belay? Belay on;* etc.)
7. Upon command, the ground operator activates the device to pull on the rope and raise the load, or to lower the load, or to stop the device, as required.

It bears repeating that effective communication is essential to ensure safety and efficiency during this type of operation.

12-6 ADDITIONAL CONSIDERATIONS

Powered equipment is no substitute for training or skill in effective rope access techniques. This equipment should be used only by competent rope access technicians who have the skill to remove themselves from the device and ascend and descend manually in the event that this should become necessary.

Powered devices are usually most effective for ascending and descending. Depending on circumstances and technician skill, these may not be the best choice where extensive maneuvers, such as rope-to-rope transfers, knot passes, or other midline actions are required. Care should be taken to avoid running knots through the device, and to avoid exposing the equipment to directions of pull that are outside the manufacturer's recommendations.

Technicians should always be familiar with and comfortable using their equipment before taking it to height. This is particularly true of powered devices. The technician must be capable of using the equipment properly while keeping hands, feet, clothing, and other equipment away from the moving parts of the device. Any time powered equipment is used, care should be taken to avoid contact with surfaces that may become hot or impose other hazards. Especially during extended use, some parts of the device (such as the engine and the capstan) could heat up. The device should be handled with care to avoid injury.

12-7 CARE AND MAINTENANCE

As with any piece of equipment used for life safety purposes, a powered rope access device should be visually inspected before use to identify wear, damage, or breakage. The device should only be used if it is found to be free of any wear or damage that would negatively influence its performance.

Manufacturer instructions should provide guidance on inspection procedures; however, some inspection elements are common to most units. For example, all approved load-bearing suspension points should be inspected for damage or wear. Cracks and grooves worn deeply in load bearing elements are strong indications of damage. Any device exhibiting these should not be used.

Similarly, all rope paths should be inspected for excessive wear. Over time, capstans and rope protection/guidance sleeves may become shiny or even grooved with wear. This could affect the ability of the device to withstand load without slippage.

Proper function of all control mechanisms should be checked and visual inspection should be carried out. Handles, buttons, levers, and rope grabs should operate properly, without sticking or moving too easily. Any moving parts of the device should be checked for proper operation. The rope drum is an especially important part of the device that should be checked to verify that the drum rotates in the direction it should and not in the opposite direction.

Finally, the overall structural integrity of the powered rope access device should be verified to ensure that no part of the unit is bent or malformed in any way. All screws should be intact, and all parts should be in good condition and aligned properly. Power cords, fuel lids, and other parts of the device should be intact and in good condition.

Some manufacturers may offer replacement of certain parts in the event that they become worn or defective. The device manufacturer should be contacted for details, and no attempt should be made to replace or modify any component without explicit involvement of the manufacturer.

12-8 SUMMARY

Powered equipment can greatly enhance the efficiency and reduce the physical effort required of rope access technicians. Such tools can be very effective in assisting the technician to ascend a rope using the engine and to descend using passive friction methods, and may be effectively used along with other methods of protection, such as positioning, as illustrated in Figure 12-8. However, powered devices should only be used by trained, certified, and competent rope access technicians whose training and skills encompass use of both powered and nonpowered techniques.

Powered devices must always be used in combination with a secondary backup system that meets the regulatory requirements of the jurisdiction(s) in which the work is being performed. Managing the backup system while using the device requires skill and practice, and the employer must ensure that any technician who is placed in a work environment is fully capable of performing the required skills safely.

The hazard analysis for the jobsite should include information that is specific and relative to the powered equipment being used, and should address such considerations as fall protection, ventilation, entanglement, hot surfaces, and other applicable matters. Rescue is always a concern, and a clear plan for personal escape (self-rescue) as well as assisted rescue should be in place prior to the equipment being deployed (Figure 12-9).

FIGURE 12-8
Powered devices can be used effectively with other methods of protection, such as positioning. Courtesy of TranSystems Corporation.

FIGURE 12-9
Planning for use of powered devices must take into consideration the potential need for rescue. Courtesy of Sean Cogan.

As always, it is the responsibility of the employer and of the competent technician to be aware of current regulatory requirements as well as hazards specific to the work at hand, and to manage these appropriately.

CHAPTER 13

Rescue

Tom Wood

By the end of this chapter, you should expect to understand:

- The importance of preplanning for rescue
- Why self-rescue is the best option for rescue
- The basics of coworker-assisted rescue
- Why a noncommittal rescue might be a safer alternative than other types of coworker-assisted rescue
- Key points for rescue from descent
- Key points for rescue from ascent
- How to prepare and train for challenging rescues
- Why standby rescue might be needed
- The difference between professional rescue and coworker-assisted rescue
- The importance of rescue refresher training

13-1 ROPE ACCESS AND RESCUE

To the layperson, the very mention of the word "rescue" might conjure up risky, adrenaline-fueled, fast-paced action and bravado, especially when applied to rope rescue. However, in reality, well-planned and safely executed rope access rescues should look more like just business as usual to the observer. Rope rescue and rope access go hand in hand. Indeed, if one compares a rope access rescue to the construction of the kernmantle ropes used in rope access, rescue could be thought of as a core (kern) skill that is surrounded by the peripheral (mantle) skills of rigging and associated technical skills. Therefore, when it comes to the rescue of a technician, rigging and rope access skills are all intertwined.

Rope access technicians can get to places others can't, such as the instances shown in Figure 13-1.

Professional Rope Access: A Guide To Working Safely at Height, First Edition. Loui McCurley.
© 2016 John Wiley & Sons, Inc. Published 2016 by John Wiley & Sons, Inc.

FIGURE 13-1
Rope access technicians often ply their trade in hard to reach locations. Courtesy of PMI.

So, while this is one of the main selling points of rope access over other means of access (e.g., fall arrest, suspended scaffolds or bucket trucks), this can also make rope access RESCUE a tricky proposition if it is not addressed before the start of work. If the possibility of a rescue from a difficult to access location is not addressed and planned for, the consequences could be tragic. While some professional or municipal rescuers may be very well prepared for any eventuality, the truth is that this varies greatly from jurisdiction to jurisdiction and in some jurisdictions responders may be untrained or ill-equipped for the rescue of a hard-to-reach rope access technician. This possibility reinforces the importance of self- or coworker-assisted rescue. In short, all properly trained and certified rope access technicians should be capable of either self-rescue or the rescue of their at-height coworkers, and of appropriately mitigating potential additional associated risks. Of course, a would-be rescuer must also be capable of recognizing where environmental or other risks are too great to allow standard coworker-assisted rescue methods, and should avoid placing themselves in danger.

From a regulatory perspective, employers of at-height workers are ultimately responsible for the safety of their workers, as well as for ensuring that there is a documented provision for the rescue of their employees. In the United States, the Occupational Safety and Health Administration (OSHA) states, "The employer shall provide for prompt rescue of employees in the event of a fall or shall assure that employees are able to rescue themselves."[1] So, while OSHA puts the responsibility for rescue on the shoulders of the employer, it does not specify how a rescue should be performed, nor does it specify what it means by "prompt rescue."

Prompt rescue for rope access technicians is important for a number of obvious reasons, but the possibility of a potential fatal condition known as *suspension intolerance* (aka *Suspension Trauma* or *Orthostatic Intolerance*) affecting a motionless, suspended technician, such as the one shown in Figure 13-2, is the primary reason for immediate action that needs to be taken.

Suspension intolerance is worth further exploration here specifically within the context of rescue. Just what rescue considerations does suspension intolerance

---

[1] Occupational Health and Safety Administration, 29 CFR 1910, General Industry Regulations.

FIGURE 13-2
Motionless suspension of a worker from fall arrest can result in the potentially fatal medical condition known as suspension intolerance in as little as 15 minutes. Courtesy of Vertical Rescue Solutions by PMI.

introduce and how can technicians protect against it when hanging suspended, awaiting a rescue?

Suspension intolerance is the body's physiological response to motionless suspension in a harness. However, the harness is not as much to be blamed as the motionless suspension, which can result in the pooling of blood in the lower extremities. According to Dr. Roger Mortimer,[2] other symptoms include a state of shock, early fainting with death, and late muscle damage. The exact physiology surrounding the causes and effects of suspension intolerance is often debated within the medical community, but one thing that everyone agrees upon is the need for the prompt rescue of a suspended worker, especially when that suspended worker is hanging from their dorsal attachment point.[3] Hanging in this manner is

---

[2] R.B. Mortimer, "Risks and Management of Prolonged Suspension in an Alpine Harness," *Wilderness and Environmental Medicine*, 22(1):77–86 (2011).
[3] *Falls From Height: A Guide to Rescue Planning*, First Edition (2013), Loui McCurley, 35–38, (2013).

not only painful, but can sometimes cause difficulty in breathing or may inhibit blood flow, especially if the harness is loose or too large for the worker. Until they are rescued, the best way conscious subjects can delay the onset of suspension intolerance is to keep moving their legs, feet, and toes until rescued.[4]

Other than treating the subject for suspension intolerance after a rescue, medical considerations during the rescue are usually limited to very basic first aid (making sure there's an unobstructed airway, observing basic cervical spine immobilization procedures, and stopping excessive blood loss). Appropriate medical treatment given by trained first responders is usually best administered on the ground and not on rope. As such, medical issues that cause or are the result of an accident are not addressed as part of rescue in this chapter.

Though it would be impossible to list all conceivable methods of rescue for the thousands of possible scenarios where a rescue could be needed, this chapter will address some common scenarios and discuss some core concepts for rope access rescues.

13-2 THE RESCUE PREPLAN

The seeds for a successful rescue plan must be planted long before an accident takes place. A well thought out, meticulously constructed and documented Rescue Preplan is a key component of the employer's Comprehensive Managed Fall Protection Program.[5] The employer's documented Rescue Preplan should be thorough enough to be of value, but not so detailed as to require a forklift to carry it or an engineering degree to understand it. Rescue tools, techniques, and resources are all identified in the Rescue Preplan.

This multilayered document, which addresses methods for the simplest rescue to the most complex and resource-intensive, includes provisions for self-rescue, coworker-assisted rescue, standby rescue (when applicable) and professional rescue. In addition, the Rescue Preplan should assign responsibilities to on-site workers, so that in the event of an emergency, everyone has a job to do. This will save valuable time by eliminating the possibility of two technicians attempting the same task during the early, hectic, and often confusing start of a rescue. Key responsibilities during a rescue include placing the initial call for help, gathering up any additional gear for rescue, directing emergency services to the rescue site and of course performing the rescue. On smaller crews, technicians may have to perform several of these duties, and having a thorough understanding of the roles everyone plays in the event of a rescue will go a long way toward ensuring a swift and successful outcome.

Perhaps most importantly, the Rescue Preplan should be shared with EVERYONE on the jobsite.

The best Rescue Preplan in the world is useless if it sits in a filing cabinet back at the home office and is not shared with those who need it most – the workers at the jobsite. (Figure 13-3)

---

[4] N.L. Turner, J.T. Wassell, R. Whisler, and J. Zweiner, "Suspension Tolerance in a Full-Body Safety Harness, and a prototype Harness Accessory," *Journal of Occupational and Environmental Hygiene*, 5(4):227–231 (2008).

[5] American National Standards Institute, *Fall Protection Code*, ANSI/ASSE Z359.2-2007.

FIGURE 13-3
The Rescue Preplan should be shared with everyone on the jobsite, usually during the daily JHA briefing. Courtesy of Vertical Rescue Solutions by PMI.

13-3 SELF-RESCUE

One of the benefits of rope access versus any other alternative means of at-height access is the ability of the technician to self-rescue. This is due in large part to simple ergonomics. When the technician is attached to their secondary safety system via their sternal attachment point (a common practice, shown in Figure 13-4) instead of the harder to reach dorsal attachment point, it is much easier and simpler to reach and unweight their safety line if it becomes loaded during an uncontrolled descent or fall.

FIGURE 13-4
The sternal attachment point is often used to attach the technician's rope grab lanyard to the safety line, facilitating self-rescue. Courtesy of Pigeon Mountain Industries, Inc.

In contrast, when at-height workers using a fall arrest system takes a fall, they are often left swinging in the breeze from their dorsal attachment, with no way to climb back to safety unless they have fallen within reach of the structure to which they are attached.

So, while the very fall arrest system they've employed has kept them from falling to their death, they are now unable to self-rescue and require a coworker to assist with their rescue. (Figure 13-5)

And then there's the equipment. (Figure 13-6) Of great benefit to the technician who needs to self-rescue, nearly all the equipment used by technicians is multifunctional. Descenders can be used for ascent, ascenders can be used for descent, and the

FIGURE 13-5
A fall arrested by fall arrest lanyards attached to the technician's dorsal attachment point can make self-rescue difficult. Vertical Rescue Solutions by PMI.

FIGURE 13-6
The multifunctional equipment use by technicians allows for vertical and lateral movement while on rope. Vertical Rescue Solutions by PMI.

mainline and safety line are interchangeable. This versatility allows technicians the flexibility to choose the best means of self-rescue, whether it means going up, down, or sometimes even laterally.

Step by step guide to self-rescue by descent (assumes mainline failure where technician is suspended from backup device at sternal attachment)

1. Verify the integrity of the remaining line. If only one rope remains, this will become the access line.
2. Attach descender to the rope, below the backup device, tensioning it as much as possible to remove slack.
3. Lock off descender according to manufacturer's instructions.
4. Attach handled ascender with etrier to the rope, above the descender.
5. Step up into the etrier and release the backup device.
6. Gently sit back down into harness, letting the descender take the load. Prevent backup device from locking up.
7. Remove backup device (Note: it is also acceptable to permit the backup device to trail above the descender during descent).
8. Descend to ground.

Step by step guide to self-rescue by ascent (assumes mainline failure where technician is suspended from backup device at sternal attachment)

1. Verify the integrity of the remaining line. If only one rope remains, this will become the access line.
2. Attach handled ascender with etrier to the rope, below backup device.
3. Connect chest ascender to the line below handled ascender.
4. Step up into the etrier and simultaneously unweight and release (do not remove) backup device.
5. Gently sit back down into the harness, letting the chest ascender take the load. Prevent backup device from locking up.
6. Remove backup device (Note: it is also acceptable to permit the backup device to trail above the descender during descent).
7. Ascend to safety.

13-4 COWORKER-ASSISTED RESCUE

If self-rescue is not feasible or possible, a coworker-assisted rescue is the next best option for a prompt on-site rescue. Certified technicians are tested on their ability to rescue their coworkers, and can usually do so with the equipment that is already a part of their kit used for daily work. In contrast with other means of access for work at height, coworker-assisted rescues form a primary component of the skills assessment process for certified technicians. For at-height workers using fall arrest, swing stages, or mechanized access, rescue training and certification is often separate from the training they receive to perform work at height.

If it is practical, the subject's main and safety lines may be used by the rescuer as long as they are not damaged, tangled, or pose an additional threat to rescuer safety. This can save the time and resources it would take to rig two additional ropes for a rescue. Any time doubts arise regarding the integrity of the subject's system, a second set of independently anchored ropes should be used.

However, no matter the method, it is safety, simplicity, and efficiency that are the guiding principles for a technician to decide which coworker-assisted rescue best suits the task at hand.

> Safety, simplicity, and efficiency are guiding principles for a technician to decide which coworker-assisted rescue best suits the task at hand.

13-5 NONCOMMITTAL RESCUE AND PRERIGGING FOR RESCUE

Noncommittal rescue refers to a rescue technique that does not expose the rescuers to further danger because they simply lower (or raise) the subject from anchors that are *prerigged* to lower (or raise) in the event of an emergency. This of course involves some preplanning, preparation, and perhaps some extra equipment. When rigging a system for a noncommittal rescue, it is important that there is enough additional rope at the anchors (not in a pile on the ground) to safely lower or raise someone without running out of rope. This is true for both the mainline and the safety line. In addition, appropriate knots should always be tied near the ends of both ropes to ensure that a technician can't be lowered (or rappel) off the ends of the ropes.

> Appropriate knots should always be tied near the ends of both ropes to ensure that a technician can't be lowered (or rappel) off the ends of the ropes.

In terms of rigging a releasable anchor to facilitate a noncommittal rescue, there are two common methods that can be used: Prerigged and locked off descenders or Munter Mule Hitches (Figure 13-7).

If using autolocking, bobbin-style descenders, they must be properly threaded and locked off per the manufacturer's instructions. It is acceptable to use descenders for the safety line as well, instead of using a fall arrest rope grab. This may make lowering easier since it eliminates the possibility of an unattended fall arrest rope grab becoming "loaded up" and requiring a short raise on the main line to release it.

If the technician chooses not to use locked off descenders at the anchor, a good alternative involves the use of releasable hitches (such as a Munter Mule hitch) tied directly to the anchors for lowering. It is important to keep in mind that while the Munter Mule hitch works well for the lowering of a stranded or injured technician, it does not work as well for a raise, and therefore extra gear may be required in that case. When using either the two descender or two Munter Mule method of prerigged systems used for noncommittal rescue, it may be necessary to have one technician operating the main line system, and another operating the safety line system (Figure 13-8).

NONCOMMITTAL RESCUE AND PRERIGGING FOR RESCUE

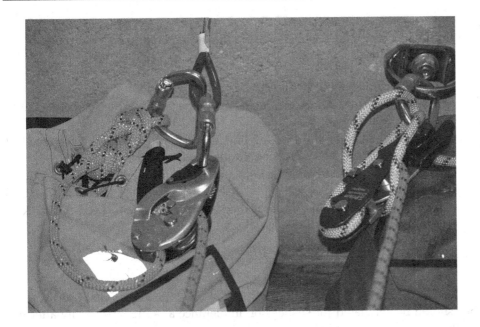

FIGURE 13-7
Prerigged and locked off autolocking descenders. Rope bags contain sufficient rope to reach the ground. Courtesy of Vertical Rescue Solutions by PMI.

FIGURE 13-8
It may work best to have two technicians lowering or raising the subject when attempting a noncommittal rescue from pre-rigged anchors. Vertical Rescue Solutions by PMI.

Steps to noncommittal rescue by lowering (assuming both primary and backup lines remain intact)

1. Make contact with the subject.
2. Ensure there is sufficient rope length to reach ground.
3. Assign one rescuer to operate the main line, and another to operate the backup line.
4. Using the prerigged descenders at the anchors, lower both lines until the subject reaches the ground.

> Steps to noncommittal rescue by raising (assuming both primary and backup lines remain intact)
>
> 1. Make contact with the subject.
> 2. Assign one rescuer to operate the main line, and another to operate the backup line.
> 3. Place a pulley on the working line, down the rope from the prerigged descender, and reeve the rope through it.
> 4. Raise the subject to safety using standard raising methods (described in Chapter 11).
> 5. Manage backup line through prerigged descender to keep the slack removed throughout the operation.

Remember that any time technicians are rescued after motionless suspension, they should be evaluated for suspension intolerance at a medical facility after the rescue.

13-6 CO-WORKER ASSISTED RESCUE FROM DESCENT

If self-rescue is not an option, the injured or stranded technician may need a little help from their friends (or at the very least their coworkers). If self-rescue or a noncommittal rescue using prerigged anchors for lowering or raising the subject isn't an option, a *pickoff rescue from descent* may be the next best method rescue technique. Most rope access organizations require the demonstration of this basic rescue as part of a technician's entry level testing process.

When attempting to rescue a technician who is injured or becomes stranded while descending, the rescuer should first alert the Emergency Medical Service (EMS) or professional rescuers that a rescue is in progress, then make certain that the scene is safe to proceed with the rescue. The next decision facing the rescuer is whether or not it's safe to use the subject's existing main and safety lines. If there is doubt about the condition of the ropes, it may be necessary to rig two additional ropes to perform the rescue. If rigging two independent ropes to access the subject, they should not be rigged to the same anchors from which the subject hangs, and must be capable of supporting a rescue load. If using the subject's existing ropes, the rescuer must be sure that their main line is the subject's safety line, and vice versa. This will allow the rescuer to access the subject on an untensioned rope.

The rescuer will then either ascend the ropes from below the subject or descend on the ropes to the subject from above, always maintaining their attachment to both the main and safety lines (Figure 13-9).

Once the rescuer reaches the subject, they should attach the subject to the rescue system with a short sling. The subject's weight should hang from the rescue system, and not from the rescuer's harness. However, it is strongly recommended that the subject has a second, unweighted backup lanyard attaching the subject to an approved attachment point on the rescuer's harness (in the event that the short sling were to fail, this second longer lanyard would take the load until the rescuer could transfer it back to the rescue system).

CO-WORKER ASSISTED RESCUE FROM DESCENT

FIGURE 13-9
It is important for the rescuer to maintain their attachment to both the main and safety lines during the entire rescue. Vertical Rescue Solutions by PMI.

Steps to companion rescue from descent by pickoff from below (assuming both primary and backup lines remain intact)

1. Make contact with the subject.
2. Ascend to the subject, using the backup line as the rescuer's access line and the subject's access line as the rescuer's backup.
3. Pass the subject as you would a knot, always maintaining a high backup device on the subject's access line.
4. Change over to descent (if applicable).
5. Connect the subject to the rescuer's descent system using a short sling.
6. Connect the subject to an approved point on the rescuer's harness, as a backup.
7. Using the subject's descender, lower them onto the rescuer's system.
8. Remove the subject's descender and original backup device from the ropes.
9. Descend to safety.

Once the rescuer has detached the subject's original descent and safety systems, the rescuer can descend, keeping in mind that if the subject was hanging motionless, he/she should be evaluated and treated for suspension intolerance at a medical facility.

Ten rope access rescue rules to live by:

1. Always alert the EMS or professional rescue that a rescue is in progress before committing resources to the rescue.
2. Real rescues are stressful; accept that and trust that your training will kick in when it's needed.
3. Rescuer safety comes first, their coworkers come second, and the subject needing rescue comes third.
4. When performing a rescue, cutting the subject free of their system should be considered a last resort, not the first choice.
5. The subject should not hang from the rescuer's harness.
6. The rescuer should always be able of escaping the rescue system without endangering the subject.
7. Treat all rescued subjects for suspension intolerance if they've been hanging motionless.
8. When it comes to prompt on-site rescue, remember that slow is smooth, and smooth is fast.
9. Improvisation during a rescue is fine, as long as you have a plan in place to deviate from.
10. During a rescue, use equipment that is appropriate for rescue loads.

13-7 RESCUE FROM ASCENT

If a technician is hanging suspended from an *ascent system* (usually a chest ascender and a handled ascender with an attached footloop on the main line) and is unable to unweight himself from that system during a rescue, the rescuer may have to build and operate a raising system to perform the rescue if a noncommittal rescue from prerigged anchors is not an option. As with the rescue from descent, the rescuer must first determine if the subject's ropes can be safely used for the rescue, or if two additional independently anchored ropes should be rigged. If using the subject's existing ropes for access by ascent from below or descent from above, the rescuer's main line should be the subject's safety line and vice-versa. If two additional ropes need to be used, they should be anchored independently of the subject's ropes, and be capable of supporting a rescue load. In addition, scene safety must be established and a call must be made to the EMS or professional rescue stating that a rescue is in progress.

Once the rescuer accesses the subject, it's time to decide what mechanical advantage system will work best to lift the subject from the ascent system (Figure 13-10). This system may be as simple as a counterbalance, or as complex as an inline 3:1 mechanical advantage (technicians will typically use their descender as the progress capture for this system, which is also known as a Z Rig).

Once the rescuer lifts the subject from the ascending system, the subject's ascenders and safety rope grab can be removed. Before descending with the subject, the rescuer may need to add additional friction to the descender to help control the two-person load (per the manufacturer's instructions).

FIGURE 13-10
Rigging an appropriate mechanical advantage system during a pickoff rescue from the subject's ascent system is an important skill to learn and practice. Vertical Rescue Solutions by PMI.

Steps to companion rescue from descent by pickoff from below (assuming both primary and backup lines remain intact)

1. Make contact with the subject.
2. Ascend to the subject, using the backup as the rescuer's access line and the subject's access line as the rescuer's backup.
3. Ascend past the subject as you would a knot, gaining sufficient height to facilitate enough space to raise the subject.
4. Change over to descent (if applicable).
5. Connect the subject to the rescuer's descent system using a short sling.
6. Connect the subject to an approved point on the rescuer's harness, as a backup.
7. Use a haul system or other raising methods to lift the subject out of their ascenders.
8. Remove the subject's ascenders from the rope.
9. Lower the subject onto the rescuer's system.
10. Descend to safety.

Once the rescuer and the subject reach the ground, suspension intolerance precautions must be taken.

13-8 CHALLENGING RESCUES

Most rope access certification organizations require entry-level certified rope access technicians to be capable of a basic coworker rescue. Usually, this rescue involves the simple lowering of the subject off their descender onto the rescuers system during a descent (aka pickoff) rescue, or performing the slightly more difficult rescue of a technician suspended from their ascent system. Steps for these have been described above.

But what if the situation at hand calls for a more involved or complicated type of rescue? Given the sometimes extreme environments and hard to reach places where technicians ply their trade, there are as many possibilities for difficult rescues as there are places that need to be accessed. The possibility of a challenging or difficult rescue is the main reason why most rope access organizations recommend that all on-rope activities happen under the direct supervision of a higher level, certified technician. In most rope access organizations, this person would be known as the Rope Access Supervisor (most often a Level 3).[6]

The Rope Access Supervisor is responsible for the on-site rescue of his or her coworkers. Though they may not have to physically carry out the rescue, they should have the knowledge and management skills to safely direct a lesser experienced technician in the rescue at hand.

Some of the more challenging rescues in rope access involve the rescue of someone from the following obstacles:

- A long rebelay
- A short rebelay
- Through knots
- Through a deviation
- From an aid traverse

Rescue from these obstacles generally involves some variation and/or combinations of the pickoff rescues from ascent and descent described above. As each rescue scenario is unique, the technician must be trained to think through each situation on its own and develop a rescue plan accordingly. Because of the wide variation in scenarios, it is not possible to address all of these in this text. Training and practice under the supervision of a competent trainer is required.

However, it's not always the obstacles that can make a rope access rescue challenging. Often, the very environment where work is being done is challenging in and of itself. High winds, extreme heights, rapidly changing weather, and remote locations all can turn what would be a relatively simple rescue in a more benign environment into an epic (Figure 13-11).

Many of the environmental concerns that would complicate a rescue are addressed in the Jobsite Hazard Analysis (JHA), which greatly reduces the risk to rescuers if appropriate mitigation methods are applied.

13-9 STANDBY RESCUE

In certain situations, the use of *standby rescue* may be the best option for a prompt rescue. Standby rescuers are usually third party subcontractors, hired by the employer to be on site, prepared to perform a rescue if the need arises. Often, this situation is called for when technicians are exposed to unique environmental hazards (e.g., bad air, possible chemical contamination, high voltage, extreme heat or cold, confined space, radiation) that pose a threat to the worker's safety, and the training and gear required for rescue is above and beyond that of most technicians (Figure 13-12).

While the use of standby rescue can sometimes be a good option for rescue, it is often cost-prohibitive, and often impractical on a regular basis for most rope access jobs.

---

[6]Society of Professional Rope Access Technicians, *Safe Practices for Rope Access Work*, August 2, 2012 edition, (6) 10–12.

FIGURE 13-11
The environment where work is performed can make even simple rescues a challenge. Vertical Rescue Solutions by PMI.

FIGURE 13-12
Standby rescuers may need to wear additional PPE to rescue technicians from a hazardous environment. Vertical Rescue Solutions by PMI.

13-10 PROFESSIONAL VERSUS COWORKER-ASSISTED RESCUE

When a rescue is above the skill level of the on-site technicians, when advanced first aid is needed, or when a rescue is simply too risky for a coworker to perform, *professional rescue* may be the best option. However, as stated earlier, the ability of technicians to reach hard to reach locations is the very thing that can make the rescue of a technician so difficult. It may well be the case that a professional rescue service is not trained or equipped to access a technician from a difficult to access location, which could delay rescue unless professional rescuers can be included in the preplan for rescue.

Even if a rescue can be carried out by the technicians on the job, professional rescue or the EMS should still be alerted before the start of a rescue. This ensures that the person being rescued can receive medical care quickly once they reach the ground, and that if the rescuer needs assistance, help is already on the way.

Once professional rescuers arrive on the scene, they will be considered the Authority Having Jurisdiction (AHJ), and will usually take charge of the

FIGURE 13-13
Including professional rescuers in the Rescue Preplan process will make for a faster rescue if they are needed. Vertical Rescue Solutions by PMI.

scene. This can sometimes create tension between the professional rescuers who arrive on the scene and the technicians on the job. A good way to mitigate this tension would be to set up a meeting with professional rescuers before work commences. Access issues, rescue procedures, lockout/tagout and specialized equipment for rescue can all be addressed during this meeting, and can go a long way toward a prompt rescue when professional rescuers are called upon (Figure 13-13).

13-11 CONCLUSION

Though rescues in rope access are relatively rare, technicians should not be lulled into a false sense of security – practicing their rescues only when it's time for recertification. Considering the potentially fatal consequences of on the job mistakes, ongoing improvements in technology and the frequency of regulatory changes, a commitment to ongoing education and rescue skills refresher training should be taken seriously by everyone who entrust both their life and their livelihood to suspended rope work.

SECTION 3

Program Administration

CHAPTER 14

Developing a Rope Access Plan

Peter Ferguson

Rope access requires a unique and disciplined approach to working at height. When successfully managed and executed, rope access can decrease accident rates, increase productivity, and cut costs. Without proper management and supervision, or without the direction of leaders who are adequately trained and experienced in rope access specific disciplines, rope access can be inefficient at best and hazardous at worst.

The remainder of this chapter is dedicated to resources and tools that will help the employer, Program Manager, and Safety Supervisor to implement good planning practices.

Clearly, a professional rope access program requires commitment on the part of the employer, but the task need not be daunting. The chapters in this section will provide simple, straightforward guidance in the form of worksheets and planning guides that may be adapted to your specific purpose. In this chapter, we will summarize and review key management aspects of rope access, while each subsequent chapter is developed around a worksheet to provide guidance on developing a specific part of the Comprehensive Managed Fall Protection Plan. Of course, it may be most appropriate to augment these worksheets with details specific to your application.

Completion of the exercises outlined in this section of the book will result in the formation of an elemental but comprehensive managed fall protection plan.

Individuals responsible for managing fall protection where rope access is used should plan and document information pertaining to at least the following:

- Policy Statement
- Work Order
- Fall Hazard Survey
- Rope Access Work Plan
- Job Hazard Analysis
- Rescue Pre-Plan
- Personnel Training Records
- Equipment Inspection and Care
- Program Audit

Professional Rope Access: A Guide To Working Safely at Height, First Edition. Loui McCurley.
© 2016 John Wiley & Sons, Inc. Published 2016 by John Wiley & Sons, Inc.

14-1 WORKING SAFELY AT HEIGHTS

Before looking specifically at the rope access plan, let's review general harness-based working at heights and consider what the minimum criteria there SHOULD be to ensure that works are carried out safely.

Rope access can be used vertically, horizontally, and anything in between, but before the actual rope works are undertaken, technicians must be able to generally work safely at heights; this is important while they are getting to the place of work and setting up their gear. While this book is about rope access, its relationship to other harness-based methods of working at height cannot be ignored.

Harness-Based Works

The concepts associated with harness-based work at height apply to situations where any worker is at height and is using a harness for safety. This would include work practices associated with fall arrest and work positioning as well as rope access.

There are three generally accepted approaches to harness-based works at height, as illustrated in Figure 14-1. Rope access sits at the top of the pyramid:

1. **Fall Arrest:** A broad term encompassing a variety of systems designed to stop a fall that has already begun. Fall arrest systems range from fall arrest lanyards with force absorbers to self-retracting lanyards to vertical or horizontal lifelines.
2. **Fall Restraint/Positioning:** The approach of securing (or tethering) a worker in a manner that prevents the person from being exposed to the fall hazard (Restraint) or to support or suspend a person at a specific location for the purpose of performing work.
3. **Rope Access:** A system of work involving certified technicians and dual rope protection that allows a person to move up, down, or horizontally by means of rope techniques.

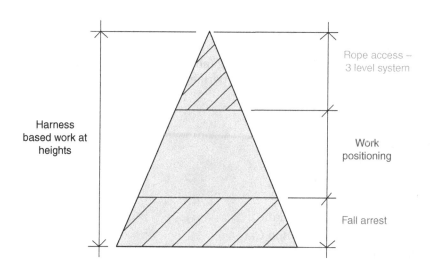

FIGURE 14-1
Pyramid of harness-based works at height.

Fundamentally, the following concepts should be prioritized whenever any of these harness-based systems of working at heights are used:

- Always ensure workers are tied-in for safety (there can be no excuse for working near an exposed edge without protection!).
- Always ensure personnel work in a team (persons working at height must NEVER work alone as there needs to be somebody present to raise the alarm and initiate rescue should an incident occur).
- Always limit (or avoid) the fall (no fall = no injury, greater fall = greater potential for injury).
- Always be prepared for a rescue – not just dial the fire brigade (this point is key – if working in a harness then the possibility of a fall and the need for a prompt rescue cannot be ignored).
- Ensure that any loading does not exceed the maximum allowable arrest force based on system type, harness attachment, and jurisdictional requirements (so all components of system will work as designed). Note that commonly accepted limits, depending on jurisdiction, are 4, 6, and 8 kN.
- Ensure that one rope is always there for the backup and as such, not loaded during normal working operations.
- A second rope (when required) may be used for provision of access or to assist the operator to position themselves so that they can work.
- Always work in a "system" where there is proper control of
  - hazards,
  - equipment,
  - personnel,
  - work methods,
  - emergencies, and
  - operator welfare.
- Always ensure that the more experienced person(s) (supervisor) oversees the inexperienced.
- Manage hazard zones and exclusion zones for safety.
- Facilitate interaction with other trades and other methods of access that may be present in the working vicinity.
- Ensure that there is adequate documentation before work begins and that "buy in" (understanding and willingness to follow the agreed documentation) by all crew is established, not just a sign-off.

While all of the above criteria are desirable traits for any work at height, in truth this does not always occur. One reason for this may be that some fall protection guidelines do not emphasize training. In fact, in some places, workers may be permitted to work at height with as little as 10 hours training in fall protection, and/or with little or no hands-on training that is specific to the use of fall protection systems. Persons with limited training are less likely to understand or perceive the importance of the fundamental requirements listed above, much less know how to implement them.

This is one advantage of using a competent and compliant rope access approach to working at height. All of these criteria are considered essential elements of even the most basic of rope access systems.

FIGURE 14-2
Making a fall arrest connection. Courtesy of Reliance Industries, LLC.

For any harness-based working at height situation, whether it is *FALL ARREST*, *WORK POSITIONING*, or *ROPE ACCESS*, there should be an entire *SYSTEM* of work. This system should take account of all of the above points to ensure that the operators are properly managed and that they remain safe while they work.

In *FALL ARREST* and often also, in *WORK POSITIONING*, this very rarely takes place.

Fall Arrest

The most commonly recognized conventional approach to work at height is, perhaps, Fall Arrest. Shown at the bottom of the pyramid of harness-based working at height in Figure 14-1, and illustrated in Figure 14-2, a fundamental – but often overlooked – concept is the realization that the operator may at some point fall, and as such, may be injured and is likely to require rescue.

This harness-based solution is common in many industries including construction, maintenance, power generation/distribution, telecom and many others. While users should be properly trained, equipped, and supervised, this is all too often not the case. Workers in these industries, particularly your entry-level craft and trades, are often poorly trained, if at all, and are frequently issued equipment that is not properly matched to the application or the conditions of the job-site. Worse yet, there is rarely any a meaningful rescue plan in place, putting fallen workers in the regrettable position of waiting for their panicked and untrained co-workers to develop an ad-hoc rescue with little or no equipment appropriate for the task. All too often, 911 is the rescue plan.

Work Positioning

The next work type in the pyramid is *WORK POSITIONING*, where the equipment is used to *position* the operator so he/she can carry out works. A typical example of this solution is a line worker on a pole where the equipment is used to lean into to assist the work – in most of these cases, the actual "access" method is a ladder or a step bolt, with the harness and pole strap providing secondary positioning and safety. This work method is illustrated in Figure 14-3.

In general, work positioning, assumes, and strives for, no falls. This is not always successfully accomplished, with many works lacking the sophistication that

FIGURE 14-3
Using work positioning. Courtesy of Pigeon Mountain Industries, Inc.

is required to cope with incidents, being properly supervised, and often lacking adequate planning.

Rope Access

Finally, there is the more sophisticated form of working known as *ROPE ACCESS*.

This solution provides a system where falls, while still possible, are highly unlikely and should they occur are likely to be of very short distance, of low impact, and can be managed quickly by the work team. Rescue capability is inherent in any rope access work team.

A properly managed rope access program will take account of all the fundamental elements mentioned above and give the client the assurance that the team can manage its operations safely and efficiently.

14-2 NECESSARY ELEMENTS OF A ROPE ACCESS PROGRAM

While there are a number of ROPE ACCESS "systems" in operation around the world, most follow internationally accepted protocols, including a 3 level hierarchical system. as found in ISO 22846.2;2012. In this system, inexperienced operators, following a rigorous basic training and evaluation, start at Level 1. As experience is gained and further training and evaluation is undertaken, a technician may graduate through Levels 2 and 3.

This hierarchical system is one of the many factors that contribute to making *ROPE ACCESS* the safe system of works that it is.

Rescue

The ability to mount a prompt rescue is critical in ANY harness-based working at heights situation. Rope access is so focused with this need that a large percentage of the operator assessments are carried out to demonstrate rescue capability.

Should an incident occur, and they rarely do, a rope access work team will have already planned for, and can quickly enact a rescue and get a disabled operator to a place of safety within minutes. For this reason, rope access teams are often called upon to provide standby rescue capability to other harness-based works.

The rescue is generally preplanned before the works begin and may be set up to remotely rescue the disabled operator (called a noncontact rescue) or alternatively, a second operator may travel to the disabled operator and then remove him/her to a place of safety (called a contact rescue). Both methods are valid and highly efficient when planned properly. A disabled operator can often be moved to a place of safety within a few minutes of an incident.

It is important that a rescue be carried out quickly as depending upon the incident (e.g., cut from a tool) the operator could be bleeding or if left suspended immobile in a harness (e.g., unconscious), which could in turn lead to further medical complications (Mortimer).

The need for rescue capability also exists at the *FALL ARREST* and *WORK POSITIONING* levels of the "pyramid," but they are very rarely planned for or practiced. In fact, the need for rescue capability particularly in the *FALL ARREST* area, where a fall is far more likely, is higher than *ROPE ACCESS*!

One Rope or Two?

In lower levels of harness-based work at heights, there will often (mostly) be only a single rope or line to an anchor used. If the operator has a method of access available, such as a roof or other elevated surface, and the line to the anchor can remain slightly loose, this is fine. As it is not required to maintain position or stability, the line is technically the fall arrest line.

In the event that the operator has to rely on the line for stability (e.g., on a sloping roof) a second line should be added for the purpose of *WORK POSITIONING*[1] so the first line can remain unloaded and be available for fall arrest.

This is rarely carried out well at the *FALL ARREST* or *WORK POSITIONING* levels.

At rope access levels, this is one of the basic tenets of the system. There are ALWAYS two points of connection such that the failure of any single system will not be catastrophic; in addition, these two systems are completely interchangeable at any given moment. It is this duality that provides the versatility required for the safety and rescue capabilities that are unique to rope access.

For nearly all works, rope access uses two ropes – rope #1 is the backup rope and rope #2 is the work rope. On occasions, works are carried out where no ropes are used and instead lanyards are used, for example, to move through a structure using what are called "aid" techniques (described in Chapter 11). In such cases, three energy absorbing lanyards are used to ensure that at least two are connected to the structure at any one time, once again ensuring "2 points of connection" at all times.

At times, it may be appropriate for a rope access technician to use conventional methods of fall protection during the course of rope access works. In such cases,

---

[1]*It should be pointed out that in some cases, and particularly so on steep pitched roofs, Horizontal Lifelines (HLL) are often specified. Advice from the manufacturer needs to be sought as these lines are generally NOT suited to loading in service and as such, a WORK POSITIONING line, if used, may need to go back to a fixed anchor point. In such cases, the use of a HLL might need to be re-considered.*

NECESSARY ELEMENTS OF A ROPE ACCESS PROGRAM 253

FIGURE 14-4
Horizontal lifeline. Courtesy of Reliance Industries, LLC.

when methods deviate from accepted rope access practices, the technician should follow all regulatory requirements and best practices for the conventional methods used (Figure 14-4).

Avoid the Fall

Rope access teaches workers to actively and always avoid a fall. Experience over some 30 years has shown this to be an amazingly effective process. A culture is built into carrying out work in a way where a fall cannot take place, but at all times, a secondary system is in place that ensures a fall, should it take place, will be short distance and impact, and can be managed by the backup system.

The rope team must always ensure they are not at risk of a fall, even while setting up, and as such, fall protection at the leading edge (such as roof edge) must be managed by the team.

Because of the effectiveness of the process, rope access teams are often called upon to carry out works that are not suspended, but on horizontal or near horizontal surfaces where they can demonstrate high levels of competence and safety while carrying out works.

Team Works

A competent rope access operator will NEVER work alone. The minimum team size is 2 persons and one of these will hold supervision skills. When the team gets large or if a number of teams are required, additional supervisors may be used to ensure the works are safely carried out and managed in a way that takes account of the site requirements including such things as working near or around other trades.

Team Documentation

Before undertaking any works, and even if documentation has been provided by others, the team should always sit down and review the work site as it presents on the day.

This is a fluid process such that several reviews may be required in a single day if the site circumstances change, and certainly at least daily while the work progresses.

The documentation must comply with the requirements of the local jurisdiction but nevertheless, must identify all foreseeable hazards, put controls in place that the team believes will address the hazards, and should also set out what form of rescue will be used for a particular incident.

In some cases a chain of command may also be identified (e.g., large sites with multiple teams where individual supervisors for each team may be used, but an overall supervisor/coordinator may also be required).

14-3 WORK IN A SYSTEM

A basic premise of rope access, as set out in the International Rope Access Standard (ISO 22846.2;2012) is that there are several key elements that MUST be managed on any rope access site to ensure a safe place of work. These are, the proper management of:

- Hazards
- Personnel
- Work methods
- Equipment
- Incidents/emergencies, and
- Welfare of operators.

Following these guidelines ensures that there is a complete system of works and these same principles may usefully be applied to the lower levels of harness-based works as well.

This *SYSTEMISED* approach attempts to take account of all relevant factors such that the team works in a manner that ensures they have adequate competency, are properly supervised, use appropriate equipment and methods, and manage all likely hazards that may present during the works. Finally, the system ensures that any necessary rescue will be swift and effective and that any issues of crew welfare are taken into account; these may be numerous issues, but typically may cover suitable mess facilities, suitable clothing for the weather and works being undertaken, etc.

While it is desirable that other forms of harness-based working at heights should also follow the above principles, unfortunately this rarely takes place.

As a result, there are many times where a client looks for a way to overlay a *SYSTEM* onto works that are taking place on a site and to ensure that rescue capability is provided – in many of these cases, rope access is the answer!

Supervisor

Rope access ensures that only experienced and suitably qualified persons act as supervisors and that every work team is controlled by an on-the-job supervisor.

Supervisors are identified based upon the hours they have spent on the job, wide range of experience, and training/assessment that ensures that supervisors are well suited to the task. It is their responsibility to ensure that those working under their direction do so safely, in accordance with agreed procedures and in a manner that will keep rescue capability as a primary goal.

The supervisor must also take account of external issues such as other trades working in the same area or the actual works being undertaken by the rope access team, which may have nothing to do with rope access (e.g., welding, cutting, concrete repairs, cleaning, painting, rigging works, etc.) As such, in addition to rope access works, the tasks the team are undertaking must also be effectively managed, which may require additional trade skills, hazard protections, public protection and the like.

Finally, the supervisor must have authority from the company to stop works immediately should he/she judge the safety of the operators or others to be compromised.

Suitable Management

The prime motive in the planning and management of rope access works is to ensure maximum safety and minimal risk.

The ISO Rope Access Standard (ISO 22846.2;2012) sets in place a requirement that a senior person within the management of the rope access company must hold minimum skills and knowledge of rope access so that an adequate understanding is in place such that the supervisor and crew is not being tasked with unreasonable or unsafe projects. This is a key factor to ensure that the crew is not placed in an unreasonable position where they feel that they are made to carry out unsafe works.

This minimum knowledge of rope access is very important also to ensure that communications between the supervisor(s) and management is effective and clear.

To this end, the work plan for a given site should be documented prior to the start of any work, reviewed periodically, and communicated effectively to ensure that all stakeholders understand

1. the project plan;
2. staffing, roles, responsibilities, and authority;
3. policies and procedures;
4. employer guidelines;
5. the protocols for hazardous materials, plant, tools or environmental hazards;
6. protections for the worksite, workers, public and other relevant matters;
7. a documented risk assessment;
8. communication plans;
9. staff competence, equipment and inspections; and
10. rescue plans and equipment.

Those assigned to supervise rope access works must have the experience and competence to supervise the task, the access, and any potential rescue for each particular rope access project under their supervision. The supervisor should be well advised regarding the skill levels of those whom they are supervising and should verify in advance the competence of all operators to ensure that they are adequately capable for the work they are expected to perform. Additional specialised information or supervision may be required in instances where specialised tools, equipment, non-rope access tasks etc. are being carried out.

During the course of work, all team members, irrespective of their seniority or experience, should "buddy check" each other's equipment; for example, check that knots are tied correctly, harnesses are buckled and adjusted correctly, and connectors are closed and locked correctly.

SUMMARY

Rope Access is a specialized mode of access and protection that is an excellent choice for working safely at height. Its proper use permits specially trained, certified technicians to access hard-to-reach places in a variety of industries, but a combination of competent certified technicians, a documented preplan, and adequate supervision are always necessary. Rope access methods are often combined with conventional fall protection methods (restraint, positioning, arrest) to accomplish engineering inspections, maintenance work, installations, and other tasks in an efficient and comprehensive manner.

Rope access technicians rely on two ropes for access and protection; a primary line for ascending, descending, and traversing, and a backup line for safety. In this type of system, the fall protection offered by the second line is completely separate and independent from the primary means of support. The two rope systems are independent yet interchangeable, offering great versatility while maintaining continuous redundancy.

The likelihood of a fall is slim in rope access, but should a fall occur, the system will ensure minimal fall distance, minimal impact force, and ability for prompt rescue.

CHAPTER 15

Developing a Policy Statement

Good fall protection practice stems from a larger corporate commitment to safety. Guidance toward developing a Comprehensive Managed Fall Protection Program may be found in ANSI Z359.2, and there also books and other resources (McCurley, 2013[1]) to help employers in this area. A good Policy Statement is considered to be a starting point for the Comprehensive Managed Fall Protection Program.

The Fall Protection Policy Statement provides general goals and guidance for a managed fall protection program (Figure 15-1). The Rope Access Policy Statement delves a little deeper, focusing specifically on a management's commitment to the rope access program as well as to providing adequate supervision, equipment, and training as needed.

Figure 15-2 shows a sample Policy Statement for a company with a rope access program as part of its Comprehensive Managed Fall Protection Program. An organization's Policy Statement may be more specific and detailed than this sample. It should not be generic to multiple companies, but should be customized to each employer and should reflect information as it is specific to the organization.

15-1 QUESTIONS TO CONSIDER

The Policy Statement is a collaborative process that begins by answering each of the following questions. NOTE: The employer, safety professionals, and workers may be interviewed to gather information by which to customize this statement for a given organization.

---

[1]McCurley, L. (ed) (2013) Index, in Falls from Height: A Guide to Rescue Planning, John Wiley & Sons, Inc., Hoboken, NJ, USA. doi: 10.1002/9781118640999.index

---

Professional Rope Access: A Guide To Working Safely at Height, First Edition. Loui McCurley.
© 2016 John Wiley & Sons, Inc. Published 2016 by John Wiley & Sons, Inc.

FIGURE 15-1
Policy Statement worksheet.

```
Fall Protection Policy Statement
       Company Name:_____
       Department:_____
Insert Policy statement in the space below:
┌─────────────────────────────────────────────────────────┐
│                                                         │
│                                                         │
│                                                         │
│                                                         │
└─────────────────────────────────────────────────────────┘

       Employer Designated Program Administrator:

              Name:_____
              Title:_____
              Company:_____
              Address:_____
              City, State:_____
              Phone_(_____)_____-_____
              Email_____

┌──────────────────────────────┬──────────────────────────────┐
│ Fall Protection Plan Prepared by: │ Rescue Plan Prepared by:     │
│                              │                              │
│ Name: _____│ Name: _____│
│ Title: _____│ Title: _____│
│ Company:_____│ Company:_____│
│ Address:_____│ Address:_____│
│ City, State:_____│ City, State:_____│
│ Phone_(_____)_____-_____│ Phone_(_____)_____-_____│
│ Email_____│ Email_____│
└──────────────────────────────┴──────────────────────────────┘
This plan is valid from _____ through _____ or until replaced by a more current plan, whichever
occurs first. This plan should be updated whenever work practices, equipment, or other aspects of the program change.
```

Use the space provided, or a separate notebook, to record your answers.

1. **List three words that characterize the employer's general philosophy about fall protection:**

   o _____
   o _____
   o _____

2. **Who is required to use fall protection?** In most cases it makes sense to ensure that anyone who works for the company, whether employee or contractor, uses some form of fall protection if they are working at height. While this may seem to be an obvious assumption to make, assumptions are not good practice. State the intent clearly.

   o _____
   o _____

> **Rope Access Policy Statement**
>
> _____(COMPANY NAME)_____ strives to protect the health and safety of all employees. As a part of this policy, Employees are required to utilize adequate fall protection whenever exposed to a foreseeable fall hazard.
>
> All employees will be trained at an awareness level to recognize and avoid fall hazards. Employees who are likely to be exposed to fall hazards will be trained and equipped to use conventional fall protection methods. Those who are specifically assigned to use rope access will be trained, certified, and equipped as rope access technicians in accordance with the requirements of the Society of Professional Rope Access Technicians (SPRAT).
>
> To support and encourage safe work at height, _____(COMPANY NAME)_____ will implement a professional rope access program into our Comprehensive Managed Fall Protection program. In keeping with good practice _____(COMPANY NAME)_____ 's rope access program will emphasize the use of adequate supervision, appropriate equipment, and certified technicians.
>
> _____(COMPANY NAME)_____ also maintains an active program to make available self-rescue, assisted rescue, and professional rescue in the event of a fall.

FIGURE 15-2
Sample Policy Statement.

3. **Who is responsible for fulfilling fall protection oversight responsibilities within the company?** Most companies have a designated safety manager; in addition, best practices in rope access dictate that one person be designated as the Rope Access Program Manager. This may be the same person as the company Safety Manager, or may be their designee. This person should be given the responsibility and the authority to manage and direct the employer's rope access program. However, it is also acceptable that this area of responsibility be under the governance of a committee or group.
    - Corporate Safety Manager: _____
    - Rope Access Program Manager: _____
    - Designated Qualified Person: _____

4. **Which types of fall protection are acceptable for use by company employees?** Identify up-front whether employees will use passive or active systems, and which type of active system(s) will be used. Active systems include positioning, restraint, fall arrest, and rope access.
    - Passive systems to be used_____

○ Active systems to be used _____

5. **Who is responsible for providing the resources employees will use when working at height?** While all employees must be charged with taking a measure of responsibility for their own safety, providing the necessary resources, such as training and equipment, is generally considered to be the responsibility of the employer. In the case of rope access, it would be the employer's responsibility to provide resources for the development, implementation, and operation of the rope access program.
   ○ Resources to be provided by employer: _____

   ○ Resources to be provided by employee: _____

   ○ Additional resources provided by _____

6. **Who will be authorized to supervise work at height?** While it takes special knowledge and training to do a good job of program management, it also takes a special skillset to manage employees on a worksite. The employer must ensure that anyone placed in a direct supervisory position is capable in terms of knowledge, training, and experience, to provide such oversight. Later in this process, the work at height supervisor(s) will be listed by name, but for now simply categorizing their qualification requirements will suffice.
   ○ Work at Height Supervisor(s) requirements _____

7. **Who will be authorized to supervise rope access work?** Within the context of work at height, supervision is an essential element that requires active participation by a supervisor trained and experienced specifically in rope access work. If this is not the same person who oversees all work at height, these two must at least work harmoniously. Later in this process, the rope access supervisor(s) will be listed by name, but for now simply categorizing their qualification requirements will suffice.
   ○ Rope Access Supervisor(s) requirements: _____

8. **Who will be authorized to perform work requiring conventional methods, and who will be authorized to perform rope access work?** While it takes special knowledge and training to perform work at height that requires fall protection, it is up to the employer to determine what minimum skillset(s) workers must have. Likewise, rope access is a unique and special type of system, and generally requires a more advanced skillset, greater training, and recognized certification. Later in this process, those individuals will be listed by name, but for now simply categorizing the qualifications of those persons will suffice.

- Fall Protection Worker(s) requirements: _____

- Rope Access Worker(s) requirements: _____

9. **To what level of knowledge will employees be trained?** Employers must ensure that employees maintain the knowledge and training necessary to safely perform the work to which they are assigned. Different employees may be exposed differently to various fall hazards, thereby warranting different levels of training and procedural guidelines. The employer can facilitate safer practices by developing and maintaining written procedures for fall protection, rope access, and rescue, thereby providing the employee with clearer guidance. This also helps to clarify and narrow the field of persons who might require additional training and/or external certification as rope access technicians.
    - Baseline for all workers _____

    - Workers using Conventional Methods _____

    - Workers using Rope Access _____

10. **What guiding document(s) will be used to determine how work is performed?** Employees should be aware of and familiar with all the policies or guidelines used to direct how work at height is performed. These should be clearly identified both for the protection of the employer and for the benefit of the employee. _____

11. **What happens when the Policy Statement needs to be changed?** Policy statements, like any other work practice, must evolve to fit the changing landscape and needs of the company and its employees. It should be made clear, however, that change requires process, and that process should be identified.

12. **What happens in the event of deviation from the Policy Statement?** While a Policy Statement should provide overall guidance, it must be recognized that the real world is filled with real challenges, and sometimes things like Policy Statements, fall protection plans, and rope access guidance documents, need to be changed. For the protection of the company, supervisor(s) and employee(s), this concept should be addressed in advance.

15-2 PUTTING IT ALL TOGETHER

Now, use the following as a guide to write your company's statement. Fill in the blanks as noted using the answers to the questions in the previous section. You may write in the spaces provided in the book, or use a separate piece of paper or computer. Feel free to modify the verbiage here as appropriate. If appropriate, include more detailed information on which employees will be protected, what constitutes an appropriate fall protection system, and in what situations fall hazards are known to exist in your organization.

The policy of _(your company name)_____ regarding fall protection is guided by the principle(s) of _(answer(s) to question #1)_____ _____. In this company _(answer(s) to question #2) _____ will be required to use fall protection. The fall protection program of _(your company name here)_____ will function under the authority of _(answer to question #3)_____. The methodologies of _(answer(s) to question #4)_____ are considered acceptable for use within this company. Resources for which _(Company name)_____ will be responsible include _(answer(s) to question 5)_____. Other resources, including _(answer(s) to question 6_____ are to be provided by _(answer(s) to question 6_____. Supervisory responsibilities for work at height will be at the discretion of _(answer(s) to question 6)_____. If/when rope access methods are used, supervision shall be provided by _(answer(s) to question 7)_____. Persons authorized to perform work using conventional fall protection methods must meet the following requirement(s): _(answer(s) to question 8)_____, and must be trained to at least the level of _(answer(s) to question 9)_____ _____.

Persons authorized to perform work using rope access methods must additionally meet the following requirements: _(answer(s) to question 8)_____ _____ and must be trained to at least _(answer(s) to question 9)_____ _____.

Work at height within _(Company Name)_____ will

be performed in compliance with _(answer(s) to question 10)_____
_____.

If changes to this Policy Statement are required, approval from _(answer(s) to question 11)_____ is required. Any actions taken outside of compliance with this statement or other company policy will be the responsibility of _(answer(s) to question 12)_____.

15-3 CONGRATULATIONS!

By completing the above exercises, you have developed a full Policy Statement that covers basic requirements.

Note that the above Policy Statement is only an example. It is not precisely the language required and should not be construed as such. In fact, based on the requirements of your organization, your Policy Statement may be quite different from this, and may require customization, changes, and/or consideration of different concepts.

Company X
Fall Protection Policy Statement
(with cross reference to guidance questions)

The policy of Company X is grounded in the principles of (1) protecting the worker, compliance with regulatory requirements, and fulfilling the needs of Company X.

It is the policy of Company X to require (2) all employees to utilize appropriate fall protection whenever they perform work at elevated locations. Methodologies used must be compliant with (4) Company X Comprehensive Managed Fall Protection Plan. This plan and may be found at //URL// and is under the direction and authority of the (3) Health and Safety Division.

It is the policy of Company X that all persons working at height shall be adequately trained to recognize and avoid fall hazards, and to use appropriate equipment and methods to protect against falls as needed. Compliance with this policy is (2) mandatory for all employees. Contractors who will be working on Company X property or on projects where Company X has authority and where employee fall protection is required shall likewise have a fall protection program that is in compliance with applicable regulatory requirements, and documentation of that program including employee training and implementation shall be available on request.

Company X recognizes the methods of (4) avoidance, restraint, and fall arrest as being acceptable means of conventional fallprotection.Use of these methods shall be consistent with applicable standards, and persons must be (8) designated as Authorized to use fall protection prior to doing so. Conventional Fall Protection will be (6) supervised by a Company X appointed Competent Person.

Company X also recognizes the use of (4) rope access as an acceptable form of non-traditional fall protection. Rope Access methods may be used (8) only by technicians who are certified to a minimum of (9) SPRAT Level 1. All Rope Access work must be (7) supervised by a Company X appointed (9) SPRAT Level 3 Technician, and must be performed in accordance with (10) Company X policy and (10) SPRAT Safe Practices Guidelines.

Regardless of the form of fall protection used, (5) Company X will select and provide the equipment used. Any personally owned equipment used on the worksite must be pre-approved by Company X and documented as such prior to use.

Any changes that diminish or reduce these minimum acceptable practices (11) shall be periodically reviewed for acceptance by the Health and Safety Division of Company X. The safety of personnel using lesser practices that are not reviewed or accepted by Company X Health and Safety Division become the (12) responsibility of the Supervisor under whose authority the work is performed.

FIGURE 15-3
Policy Statement example, with annotations.

The Policy Statement shown in Figure 15-3 shows an example of how language within the Policy Statement might read based on the use of the exercises in this chapter to improvise verbiage. Annotations are provided as reference to show where each question is addressed.

CHAPTER 16

Writing a Work Order

When an external rope access company contracts with an employer to perform rope access work on a given site, a work order may be used to provide an overview of the work to be performed and to authorize the contractor to work within the scope of the agreement. A work order may also be used within a company to communicate between departments or divisions. Typically, a work order will summarize customer information, describe the location and type of work to be performed, and provide an estimate of charges for material and labor. In some cases, the work order is also used as an invoice.

A work order may be generated by the rope access contractor and submitted to the customer, or it may be generated by the customer and submitted to the rope access contractor. Either way, the content of the work order forms the basis for creating a bridge between the customer and the contractor, so the two must work closely together to ensure that their needs are adequately identified and included in the order (Figure 16-1).

The work order should clearly identify the location of the work to be performed, as well as the scope of work to be performed. The specific methods and equipment are not necessarily included here, but these aspects should be considered as they will influence other things, such as the number of staff required and time for completion. An estimated value, taking into consideration a particular rate of pay and the total hours required, will typically be included as part of the work order.

The following questions may be used as a guide for collecting the kind of information required for a work order.

1. Customer Name:_____
   This may be the name of the company that intends to contract external resources, or it may be the name of a department that intends to use the services of another department.
2. Customer Address:_____
   _____
   _____
   The address of the customer's home/office location should be listed here, and if the billing address is different this information should also be noted.

Professional Rope Access: A Guide To Working Safely at Height, First Edition. Loui McCurley.
© 2016 John Wiley & Sons, Inc. Published 2016 by John Wiley & Sons, Inc.

Rope Access Work Order

Date:

Customer Name:________________ Contact:________________

Business Address:

Billing Address:

Phone:________________ E-Mail________________

Alternate/Emergency Contact________________

Location of work:

Scope of work:

Start Date (Target) ________ (Actual)________ Completion Date (Target)________ (Actual)________

Service Provider:________________ Contact:________________

Phone:________________ E-Mail________________

Responsibilities:

Additional Requirements:

Materials Cost $________________ (Estimate)

Labor Cost: Hourly Rate ________ Number of Workers ________ Total Estimated Labor$________________

(Material or Labor Cost Variances greater than 10% Will be Pre-Approved)

AGREED AND APPROVED

Customer:

________________ ________________ ________________

(Sign) (Print) (Date)

Service Provider:

________________ ________________ ________________

(Sign) (Print) (Date)

FIGURE 16-1
Sample of rope access work order.

3. Point of Contact:________________________________

   ____

   ____

   This is the name of the specific individual representing the customer who will serve as the point of contact for the work to be performed. The name, telephone number, email address, and alternate contact (if applicable) are useful information to have.

4. Location of work:________________________________

   ____

   An address, GPS location, or other physical description of the work location is an important piece of information. It should never be assumed that the work to be performed is at the same location as the main company headquarters. The very nature of rope access makes it conducive to being used in remote locations, some of which may not have a physical address.

The information provided here should be thorough enough to guide the reader to the jobsite.

5. Rope Access Provider Name:_____
   This may be the name of an external subcontractor that will provide rope access services, or it may be the name of a department that will perform rope access services for another department.

6. Rope Access Provider Address:_____
   The address of the rope access provider's home/office location should be listed here. Of course, if the provider is simply a different department within the same organization, this address may be the same as that identified in question #2.

7. Rope Access Provider Contact:_____

   This is the name of the specific individual representing the rope access services provider, and is the one who will serve as the point of contact while work is being performed. The name, telephone number, email address, and alternate contact (if applicable) are useful information to have.

8. Work Start Date (Possible): _____
   The earliest date on which work may be performed should be shown here. This does not necessarily mean that work WILL start on this date, it is simply to identify when work MAY begin. This date should be inserted early in the process to provide a parameter for the provider, whereas the date noted in question #9 may need to be adjusted based on circumstances.

9. Work Start Date (Actual):_____
   This is the date on which work actually begins. This date may need to be updated during the course of the project.

10. Work Completion Date (Required):_____
    This date reflects the absolute deadline for completion of the work. The target for completion should be earlier than this date, if possible, but the required completion date provides an important planning parameter for the service provider.

11. Work Completion Date (Actual):_____
    This is the date on which work is actually completed. The goal for this date should be no later than the date listed in question #10, and if possible the target should be set earlier. At first, this space should reflect the target completion date but it may need to be somewhat flexible depending on the nature of the work and the people involved who influence the work. Actual completion date should be updated if necessary.

12. Service Required: _____
    _____
    _____
    _____
    _____

    This section should thoroughly describe the services to be performed and the goals to be achieved for the work to be considered "finished."

13. Rope Access Service Provider Responsibilities:_____
    _____
    _____

    It is very important to clearly identify what aspects of the work the rope access provider is responsible for providing. Explicitly noting what kind of labor, materials, equipment, supplies, transportation, or other services are required of the service provider will assist in planning and will prevent misunderstandings later.

14. Additional Requirements:_____
_____
_____

If the customer has any additional requirements, these should be noted here. Some examples of additional requirements might include requirements for security clearances, certification requirement, proof of insurance, coordination with other personnel or resources, or other company-specific prerequisites.

15. Cost Estimate – Materials _____
Cost Estimate – Labor _____

An experienced rope access program manager should be able to at least approximate the cost of equipment and time to do a job. Materials will include any tools, materials, supplies, or additional rope access equipment needed to accomplish the assignment. Labor must include all facets of planning, supervision, work, and inspection, and therefore may require separate rates for multiple lines. Of course, it is not always necessary to break the information down in such detail for the customer – this is left to the program manager's discretion.

16. Authorization Signatures: This completed form should be signed by representatives of both parties to provide authorization for the work to be performed.

SUMMARY

A work order is a potentially useful document for coordinating rope access services and ensuring clear understanding between parties. The example information provided in this chapter should be considered a starting point, and additional information should be added as needed.

CHAPTER 17

Establishing a Work Plan

Many employers understand fall protection in conventional terms, and decades of experience has resulted in common familiarity with such methods such as guardrails, restraint systems, and fall arrest. When an employer chooses to implement a rope access program into their fall protection program, acknowledging the place of rope access as part of that program helps to establish a good framework. In this chapter, guidance specific to developing a rope access work plan is provided. This information should be considered only as a starting point, as additional information may be required for any given workplace.

A Rope Access Work Plan, also sometimes called a Rope Access Permit, is a document that is created and maintained by the employer to provide guidance and clarification regarding the scope within which rope access technicians are authorized to work. In most cases, the organization should maintain an over-arching Work Plan on file, which provides general guidance, and will complement an additional specific work plan for individual projects.

Developing an individual plan specific to a work is important because rope access is not a "one-size-fits-all" solution. The reason that all rope access technicians must be trained in a variety of specific techniques and skills is because a range of different techniques, equipment, and even skillsets may be required for any given job, depending on the work environment (Figure 17-1).

The Rope Access Work Plan should take into consideration the contents of the Job Hazard Analysis (Chapter 18) and Fall Hazard Survey (Chapter 19) documents, but is intended to reach further than either of these. This document delves deeper into the specifics of the rope access equipment and methodologies, including those of how the work is to be performed, the equipment to be used, and the safety provision for both workers and bystanders. The Work Plan should be shared with all affected personnel prior to the start of work.

At least the following information should be considered and included in the Rope Access Work Plan.

1. Project Name or ID

Every project should have a work plan associated with it. It is acceptable to create an over-arching Rope Access Work Plan that may be used for different jobs where work is essentially the same, but consideration should be given to the applicability of the plan for each site before it is used.

Professional Rope Access: A Guide To Working Safely at Height, First Edition. Loui McCurley.
© 2016 John Wiley & Sons, Inc. Published 2016 by John Wiley & Sons, Inc.

FIGURE 17-1
Sample Rope Access Permit.

Rope Access Permit
(Work Plan)

Date Prepared_____   Date(s) of Work_____
Company Name_____Project_____
Preparer's Name_____Title_____

Location of Work:

Scope of Work:

Access Methods
☐ Ascending   ☐ Long Re-Anchor   ☐ Rope-to-Rope   ☐ Other_____
☐ Descending   ☐ Short Re-Anchor ☐ Traverse   ☐ Other_____
☐ Knot Pass   ☐ Deviation   ☐ Lead Climb   ☐ Other_____

Anchorage Description:

PPE/Rope Access Equipment Required:
☐ Harness   ☐ Descender   ☐ Carabiner #___   ☐ Steel Toe Boot
☐ Helmet   ☐ Chest Ascender   ☐ Pulley #___   ☐ Other:
☐ Gloves   ☐ Handle Ascender #__ ☐ Workseat   ☐ Other:
☐ Eye Protection   ☐ Backup Device(s) #__☐ Powered Device   ☐ Other:
☐ Hearing   ☐ Cow's Tails #___   ☐ FR Clothing   ☐ Other:

Work tools to be used:

Work Team Members:
　Supervisor:_____ Certification:_____ Duties:_____
　Worker:_____ Certification:_____ Duties:_____
　Worker:_____ Certification:_____ Duties:_____
　Worker:_____ Certification:_____ Duties:_____

Public Safety Provisions:

Rescue Plan (check all that apply)
☐ Self Rescue
☐ Co-Worker Assist   EMS Response Contact   Nearest Medical Facility
☐ Standby (Onsite)
☐ Municipal Response

Other Relevant Documentation to be attached, including Work Order, Job Hazard Analysis, Fall Hazard Survey, etc.
Use Additional Pages as necessary

Relevant information for a given project should be collected at a single location for future reference. This can be helpful during the project as well as for later, should questions arise. The practical employer will also begin to recognize similarities and differences between projects over time, and can use collected information as a learning tool for themselves and the field technicians.

2. Work Location

Even where a single rope access plan is used for different but similar jobs, the plan must be re-evaluated with consideration to each work location and scope of work to ensure that it is applicable and that no changes need to be made. Even where work is very similar to previous projects, environmental conditions can change requirements and considerations. Adjacent work, nearby facilities, and even the time of day can also have an effect.

Every work location should be evaluated on its own merits, and assumptions should not be made about a site based on previous evaluations of either that particular site or similar sites. Recent modifications and repairs, new installations, and changes to the facility may require a revision of the rope access plan in terms of anchorages, exposure, leading edges, and other factors.

3. Rope access methods

All rope access technicians should be trained and certified as being capable of performing all fundamental skills at their level of certification. That said, not every technique is required for every job. Listing the methods that are expected to be used in the work plan will help to ensure adequate preparation for the job, but this does not necessarily preclude the need for changes to the plan or for additional skills that may be required.

Rope access methods should be selected based on the work to be performed as well as site conditions. Methods that may have been used previously on a site should be re-evaluated when new work is to be performed on that site to ensure that the methods are appropriate to the site in its current state as well as appropriate for use with the work and tools to be used on the current project.

4. Members of the work team

Each member of the work team should be listed by name, and his/her duties should be listed, and verification should be made to ensure that the required skills are commensurate with each person's skill/certification level. It is the responsibility of the rope access supervisor to assess the individual team member's suitability for the work to be performed.

It is not only the skills of the individual that matter, but also the ability of the individual to work collectively with other team members. Team members who are not familiar with one another should be provided ample opportunity to establish rapport and working relationships with one another before having to face difficult challenges together.

5. Equipment

While the basic equipment carried by most technicians is quite similar, identifying the specific equipment that will be used helps to set clear expectations and prevent surprises.

Part of selecting the appropriate equipment involves the evaluation and determination of which standards apply to which equipment. Because only few standards that are specific to rope access equipment exist, some equipment may not be able to be labeled with an appropriate reference standard. In such cases, it is the responsibility of the employer to be knowledgeable about performance requirements and to verify the suitability of the equipment. Conventional "fall protection" equipment that is rated to be compliant for fall arrest may not be appropriate for rope access work due to different construction and performance requirements. It is essential to ensure that any equipment used is appropriate for the purpose for which it is being used.

6. Hazards associated with the work to be performed (JSA, JHA)

All personnel should be familiar with identified hazards, and should be permitted to provide input to this information as appropriate. The Job Hazard Analysis, discussed in Chapter 19, is an important component of the Rope Access Work Plan.

Hazards specific to the use of rope access should also be taken into consideration as part of this analysis, and verification made that the rope access is an appropriate means of access for the work. The employer should confirm that the suspended person will be able to safely use materials, equipment, and tools, and that the work tools themselves will not place the worker at undue risk. Care should also be taken to note whether the work may loosen material which could become a hazard to the worker or others, or whether the time required for the work at any one location is acceptable and is within the comfort level of the exposed worker(s).

7. Personal Protective Equipment (PPE)

PPE, discussed in Chapter 5, may vary somewhat depending on the work to be undertaken. Requirements for PPE should be established by the employer and the importance of using appropriate PPE should be emphasized to all workers. The employer should ensure that PPE provided to workers fits them properly and is appropriate for the task.

It is the responsibility of the site Supervisor to monitor and ensure that the appropriate PPE is being properly used.

8. Anchorage Security

The importance of security of system anchorages cannot be overstated. Anchors should be rigged to incorporate appropriate safety factors. Because anchors are a fundamental element of any rope access system, the attachment to the anchorage should at least equal the strength of the system attached to it. Re-direction of the ropes from an anchor should not exceed 120° unless the side loads produced at the redirection point are considered. Similarly, where the included angle at the attachment is high and produces a "multiplier" effect, the extra forces produced should be considered.

Establishing a safe and appropriate anchorage and connecting the system properly to that anchorage is just the beginning of maintaining anchorage safety. With the potential for technicians to be on-rope and out of visual contact with anchorages, some provision must be made for ensuring that the anchorage is adequately monitored for consistency, and that it is not tampered with. If anchorages are left in place between work shifts, a safety check should be made before the anchor is re-used by the next shift.

9. Public Safety

Where rope access work takes place in areas where other trades or even the public may be exposed to hazards, provision should be made for protecting life and personal property within the hazard zone(s). It is the responsibility of those undertaking the rope access work to ensure that some means of warning and/or protection are in place.

Hazard zones should be established and marked, blockaded, or identified to warn rope access personnel and passers-by of hazards associated with the work being performed. This might involve signage, barriers, and/or even attendants to help prevent inadvertent exposure.

10. Rescue

Of course, wherever work at height takes place, a plan for prompt rescue must also be in place. Certified rope access technicians should be capable of carrying out self- and coworker-assisted rescue, but knowing how to get an individual off a rope isn't enough to be considered "rescue capability."

All aspects of a potential need for rescue (including extrication, proper packaging, trauma care, treatment for possible medical conditions, extended care, and multiple casualty incidents) should be considered and adequate preparation made. It is also prudent for any person who requires coworker-assisted rescue to be subsequently evaluated by a medical professional.

17-1 SUMMARY

The Rope Access Work Plan (Permit) is simply a written statement prepared by the employer, which describes how a particular job (or "types of jobs" where these will be essentially identical) should be undertaken. The emphasis of the plan is to ensure that any risks to the health and safety of the workers or others who may be affected, are minimized. All affected employees should be given opportunity to review and discuss the plan before commencing work. It may be useful to attach a signature page to the plan to record an acknowledgment by employees for having reviewed the plan.

CHAPTER 18

Performing a Job Hazard Analysis

A Job Hazard Analysis (JHA) (also sometimes called a Job Safety Analysis (JSA)) is a document that is created to provide a general overview of a wide range of hazards found on a jobsite. It applies to all workers on a site, not just to those who are working at height. The JHA should not be confused with the Fall Hazard Survey, which is intended to be more specific to the fall hazards identified as part of the work. These two documents are both separate but important parts of a good Rope Access Work Plan (Figure 18-1).

A "hazard" is considered to be any condition that exposes a person to the potential for harm. Regulatory authorities often associate hazards with a condition or activity that, if left uncontrolled, can result in an injury or illness to a worker. It should be the goal of every employer to identify and eliminate hazards to help prevent injuries and illnesses. This is the purpose of the JHA.

18-1 THE PROCESS

A JHA should be conducted for any job that might expose a worker to potentially severe or disabling illness or injury. Jobs where human error has a high likelihood of increasing risk are especially important to be considered for conducting the JHA. The evaluation process should focus on the relationship between the worker and the work that is to be performed and must take into consideration the task, tools, and environment, as well as the workers themselves.

Any worker who is likely to be exposed to such risks should be involved in the hazard analysis process. Being involved in the analysis process gives a person a unique and valuable perspective, and this can be very helpful in identifying hazards. Employees should be encouraged to identify risks that might exist in their surroundings, and should be involved in brainstorming sessions as well. Encouraging employees to be involved in the hazard analysis process will also help them to "own" solutions when they are proposed.

A good starting point for any hazard analysis is to review accident history. Those incidents that may have required treatment, near misses, or a specific loss, should be scrutinized closely for related hazards that can be mitigated in the future.

Professional Rope Access: A Guide To Working Safely at Height, First Edition. Loui McCurley.
© 2016 John Wiley & Sons, Inc. Published 2016 by John Wiley & Sons, Inc.

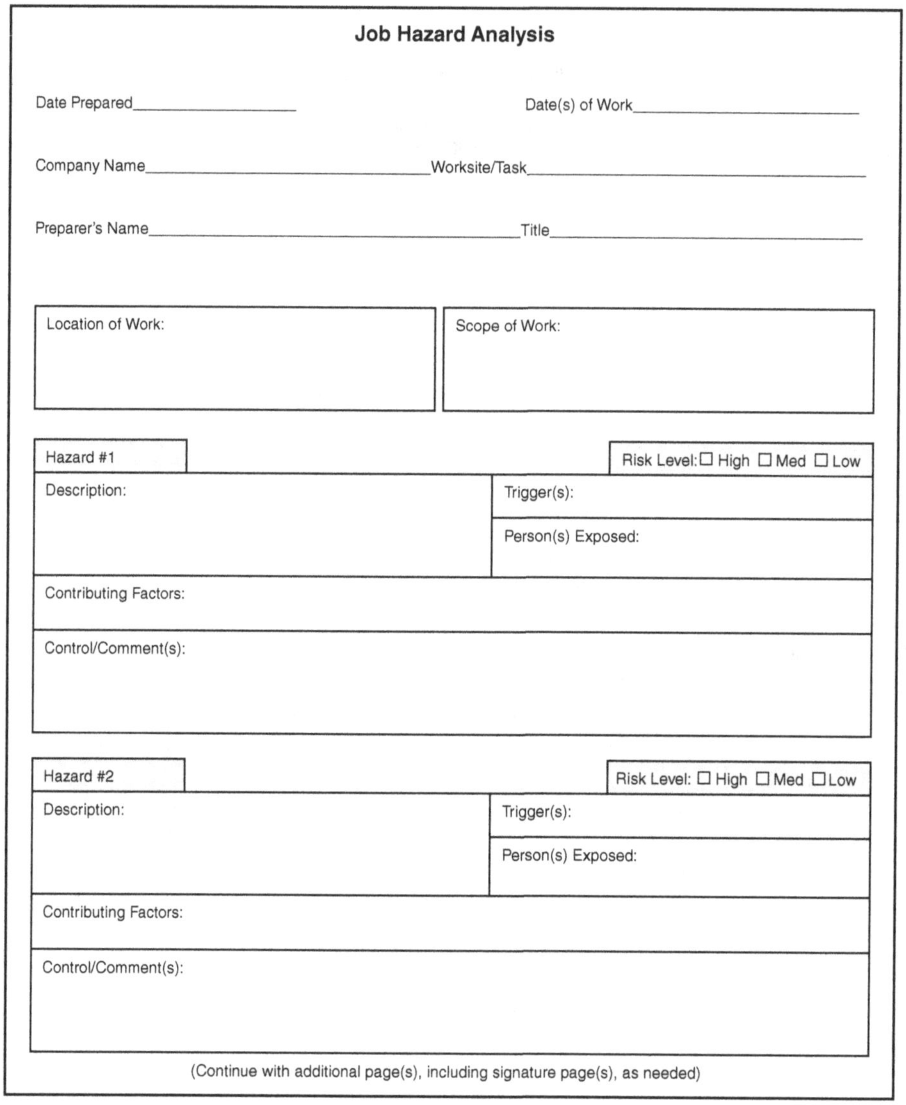

FIGURE 18-1
Sample JHA form; where applicable, provision should be made to record the acknowledgment of workers briefed with this information. A signature page works well for this.

Near misses and actual incidents are not the only indicators of hazards. Exposures that have not yet resulted in an incident or in near misses should also be identified. It may be helpful to consider specific jobs, and to break each job down into individual tasks, and then to look for particular hazards at each step. If possible, observing employees performing the tasks and recording potential hazards as they are observed will be useful. It may also be helpful to use photographs or videos for reference.

18-2 CONTENT

A good JHA will identify circumstances that may pose a risk to employees, probability of exposure, and consequence of experiencing the hazard. For each hazard, consideration should be given to the environment, exposure, trigger, consequence, and contributing factors.

Most hazards involve multiple influences that combine together to create a hazardous condition. One example of a hazardous condition in a rope access environment might be a technician at a refinery (environment) whose air quality (exposure) is compromised when a wind shift (trigger) blows exhaust from a nearby chimney (contributing factor) their way and causes them to become lightheaded and faint (consequences).

To perform an analysis of this situation, you would need to consider the location where the work is taking place, who might be placed at risk, what environmental factors might exist, what can possibly go wrong, how it could happen, and what the consequence of exposure might be. Let's use this example to complete a JHA form, as shown in Figure 18-2.

- *Hazard Description*. In this box we answer the question *"what can go wrong?"* A hazard is only a hazard if it can have a potential negative effect on the worker. In this case, a worker could be exposed to poor quality air, and this could potentially cause discomfort, disorientation, or even loss of consciousness. The description of the hazard should describe both the cause and the potential effect of the hazard.
- *Trigger*. The trigger for the hazard may involve factors that are controllable or uncontrollable. In the scenario described, the rope access team has no influence over how efficiently the gases are burning in the refinery, so although this is clearly the primary trigger it is not one that the team can control. For controllable elements, the team must look to contributing factors.
- *Contributing Factors*. Fortunately for the rope access team, air quality in an open environment is unlikely to become catastrophic very quickly. This permits some time to identify and mitigate the potential hazard. This is an important consideration, and is helpful in determining the probability and potential consequence of an incident so that appropriate hazard controls may be selected.
- *Persons Exposed*. Workers can only be protected if they (or the employer/supervisor) are aware of their potential for exposure. Here it is essential to identify who are likely to be exposed so that adequate protection can be provided.
- *Risk level*. This question involves a certain amount of judgment. It must take into consideration the likelihood that the hazard will occur, the probability of someone being exposed, and the potential consequence of exposure. If previous incidents or "near misses" have occurred, the likelihood of recurrence may be higher. A severe consequence, such as unconsciousness, would also be a parameter for a higher risk ranking. In the example provided, if the probability is considered low but the potential consequence is high, a medium ranking would be appropriate.
- *Mitigation*. Once the hazard is identified, consideration should be given to various potential methods for mitigating that hazard. Some possible solutions might include technicians carrying air monitors, monitoring of windspeed and direction onsite, shutting down the work that is creating the exhaust, or providing respiratory protection to the workers. This is not an exhaustive list, but is intended to show that there may be many possible solutions for mitigating a given hazard and that each potential solution should be weighed based on risk:benefit, and with consideration as to whether the potential solution might create additional hazards.

FIGURE 18-2
Sample completed hazard analysis.

```
                            Job Hazard Analysis

Date Prepared    June 18, 2025              Date(s) of Work   June 23–29, 2025

Company Name  PMI Solutions Rope Access   Worksite/Task    ABC Refinery

Preparer's Name     Joe Access              Title  SPRAT L3 Supervisor
```

Location of Work:	Scope of Work: Dye Penetrant Examination (PT) of welded T-joints
Splitter Tower inspection, Sector B	

Hazard #1		Risk Level: ☐ High ☒ Med ☐ Low
Description: Exposure to contaminated air could cause nausea, vomiting, dizziness, faintness, and/or shortness of breath.	Trigger(s): Inefficient combustion of waste gases	
	Person(s) Exposed: Any Technician on site	
Contributing Factors: Wind direction/speed, Proximity of technicians to emissions		
Control/Comment(s): Technicians will not work in close proximity of waste combustion processes; one technician per workteam will wear air quality monitor		

Hazard #2		Risk Level: ☐ High ☐ Med ☐ Low
Description:	Trigger(s):	
	Person(s) Exposed:	
Contributing Factors:		
Control/Comment(s):		

(Continue with additional page(s), including signature page(s), as needed)

18-3 USING THE JHA

A list of identified hazards is only useful if affected parties know about it, and understand the information it contains. It is good practice to hold a brief meeting as needed to update workers on hazards and the plans for mitigation, to answer any questions they might have, and to ensure that there are no "disconnects" between the plan and the worker's ability to execute it.

Documentation that all affected workers have participated in such a briefing and that they understand the hazards and mitigation can be readily achieved by adding a signature page to the JHA. Here all workers should sign to acknowledge their awareness prior to beginning work. This is also a good place for workers to make note of personal conditions (such as allergies) that may increase their risk on the worksite.

By following the steps in this example, hazard analysis activities can be organized.

18-4 SUMMARY

Employers and supervisors may find that a job hazard analysis is useful in helping to eliminate and prevent hazards in the workplace. The ultimate goal of any process such as this is to develop safer and more effective work methods and thus prevent worker injury and illness. A safer workplace is proven to result in reduced workers' compensation costs and increased worker productivity. In addition to contributing to a reduction in injury and illness, a JHA can provide the foundation for training new employees to perform their jobs safely.

Many employers use a JHA as the foundation for periodic or daily safety briefings. This is generally a good practice, but in the case of rope access the JHA is but one part of the Rope Access Work Plan and so rope access employers should consider reviewing the Rope Access Work Plan in its entirety, to include the JHA, at such safety briefings. Providing an acknowledgment page for employees to sign at the briefing can be useful in documenting participation.

CHAPTER 19
Fall Hazard Survey/Assessment

A Fall Hazard Survey is recommended by the ANSI Z359 process as part of any Comprehensive Managed Fall Protection Plan. One example of a Fall Hazard Survey is found in Figure 19-1. The purpose of the survey is to identify and consider potential fall hazards in the workplace.

Employees should only be permitted to work on walking-working surfaces that are structurally sound and have sufficient integrity, including strength, to support the work activity. This means that a person designated by the employer must evaluate such work surfaces to make such a determination. The number of workers, the weight of the worker(s) and any relevant equipment should be taken into consideration. On a relative note, determination must also be made as to whether fall hazards might exist at the work location. The employer should ensure that these hazards are sufficiently explored, identified, and mitigated.

19-1 CONDUCTING THE SURVEY

While it is ultimately the responsibility of the employer to ensure that the survey is conducted, normally the responsibility to conduct the survey is entrusted to the Program Manager or the Competent Person for fall protection. It is also appropriate for the survey to be conducted by a delegated resource, such as a hired consultant, a specialist, or another employee. The survey should reflect the name (and, if appropriate, contact information) of the individual who prepared it, in case any future follow-up or additional information is needed.)

A fall hazard survey should be inclusive of all potential fall hazards, but each hazard should be dealt with individually. This means that a fall hazard survey for a plant or organization may be segmented into several work locations, and within each segment multiple fall hazards may be identified. Each fall hazard that is identified as being location-specific should be given an alpha-numeric reference number to differentiate it from others. Several hazards may be identified at a given site, and each should be given individual consideration. Where several hazards of a similar type and nature are found in multiple locations, it may be appropriate to consider these together. The fall hazard survey for a given worksite may include multiple pages, and multiple fall hazard worksheets.

Professional Rope Access: A Guide To Working Safely at Height, First Edition. Loui McCurley.
© 2016 John Wiley & Sons, Inc. Published 2016 by John Wiley & Sons, Inc.

```
                        Fall Hazard Survey

Hazard ID _____
Prepared by: _____      Date_____
Is this a Location Specific or a Multi-Location Hazard? (circle one)
Describe Actual or Typical Location:
_____
_____
_____


Diagram/Detail of Hazard:



How is this location typically accessed? (consider preparing a separate Hazard ID for access)
_____

Type of Work occurring in this Location
_____
_____

Persons Potentially Exposed (include number simultaneously exposed & duration of task)
_____
_____

Height of the potential fall: _____meters (_____feet)
Obstructions:_____
_____

Additional Environmental Considerations: (check all that apply below. Use separate page to diagram/describes as needed)
        ☐ Moving equipment & materials      ☐ Sparks, flames, and heat
        ☐ Unstable, uneven & slippery       ☐ Hazardous chemicals
          walking/working surfaces          ☐ Electrical hazards
        ☐ Unguarded openings                ☐ Environmental contaminants
        ☐ Climate & weather factors         ☐ Sharp/abrasive surfaces
        ☐ Other notable factors
History of previous incidents in this (or similar) location:
```

FIGURE 19-1 Fall Hazard survey.

A wise Program Manager will ensure broad collaboration among workers, supervisors, and other affected persons during the course of a fall hazard survey. Ideally, a person who is familiar with building operations and work procedures should accompany the individual conducting the survey. The workers who perform a task regularly are the best source for information about work paths, workers' movement at the work station, limitations to mobility, frequency and duration of exposure, distractions, and other matters related to work practices. It is important to ensure an open path of communication, with the goal of collecting information about how things are really done – not just how they are supposed to be done. Past records of incidents, exposures, and near-misses are also an excellent resource for collecting information and identifying fall hazards.

19-2 SURVEY CONTENTS

It is not necessarily required to use a specific form or format such as that shown in Figure 19-1 for the survey, but the report should be written. All potential fall exposures, including unprotected edges, access and egress points, exposure at locations where workers will be standing or working for extended periods, and any other exposures should be noted.

Each fall hazard should be identified as a separate hazard, and considered individually. For every location where a fall hazards is identified, thorough notes on that specific hazard should be prepared, including details about the surroundings. Observations regarding the working surface, potential fall distance, location and distance to obstructions, available clearance, and other hazards (lockout tagout, chemical, heat, etc.,) in the area could all be important factors, so these should be documented in detail. In addition, details of how many people are likely to be exposed to the hazard at any given time, and what time(s) of day or under what circumstances they are exposed should be documented.

Additional information gathered at this time will lend itself to the next phase of the process, which is mitigation and fall protection planning. Details of whether there appear to be other ways to do the job that might expose the worker to less risk should be noted. The feasibility of these potential alternatives with the workers most familiar with the work station must be discussed. Alternative recommendations should not be made without first exploring the limitations of the alternative(s). Whether or not potential alternatives exist, details of anything that might be relevant to provision of fall protection, including edge conditions, anchorage location(s) and type and clearance distances should be noted. Even if this information is not made use of at that point of time, collecting the facts now is easier than coming back for them later.

Finally, any obvious hazards that could affect at-height workers or their equipment should be identified. This would include heat sources, hot objects, sparks, flames, electrical hazards, chemical hazards, sharp objects, abrasive surfaces, moving equipment and materials, air quality, environmental/weather factors, and other conditions. Such hazards are not always constant, but may be variable over the course of one or more days, or may only be present during certain types of work. This is another time when interviewing affected workers can be very enlightening.

Each hazard should be documented thoroughly. Sketches or photographs may accompany notes to help clarify layout and configuration of the environment.

Some means of prioritization should be used to analyze the severity and significance of each hazard. This should take into consideration both the probability of exposure, and the consequence of an incident. This can be as simple or as complex as the employer chooses. One simple example is shown in Figure 19-2.

Employers may choose to approach different levels of severity in different ways; for example, by placing higher restrictions or personnel training requirements on hazards that have a higher rating.

19-3 USING THE SURVEY

Information that is collected during the hazard survey should be documented and organized into a logical format for future reference. This information will be useful

FIGURE 19-2
Measuring Hazard significance.

Hazard Significance

A. Frequency of Exposure: 1.......2.......3.......4.......5
                              Never........................Continuous

B. Injury Potential if a fall should occur: 1.......2.......3.......4.......5
                                               None...........................Fatal

$$\frac{(A)}{\text{Frequency}} + \frac{(B)}{\text{Injury Potential}} = \frac{}{\text{Priority}}$$

in assisting in the consideration and selection of appropriate means of protection, and in updating plans for future work. A completed survey is the starting point for determining and outlining protection method(s).

Surveys should also be retained for future reference. Even after new surveys are performed it can be useful to hold on to a selection of previous surveys to help provide historical reference of information and change.

The fall hazard survey must be revisited whenever a substantial change occurs. A substantial change is any change that might impact safety in the workplace. This could include structural changes that are made to the environment, changes to work procedures, changes in adjacent work, equipment and/or personnel changes, and anything else that could influence safety. Aside from taking specific changes into consideration, the survey should also be reviewed and verified at least annually.

19-4 FALL HAZARD MITIGATION

The final step in the survey is to develop a mitigation plan. Whenever possible, elimination of the hazard should be the primary objective. When elimination of the hazard is not the best option, other control measures may be selected based on need.

Initially, ALL possible solutions should be considered. For each hazard, it should be noted whether each type of mitigation/protection is *possible*, even if that type of mitigation is unlikely to be used. Notes may be made at this time or later regarding the pros and cons of each method. The important thing here is to clearly acknowledge the possibilities. Understanding the range of possibilities is essential to determining the *best* course of action. An example of a worksheet for documenting possible mitigation/protection methods may be found in Figure 19-3.

It is important to recognize that the "best" option for a given situation may vary depending on the type of work being performed, access, egress, work procedures, and other factors. The "best" solution for one type of work at a given location may differ from the "best" solution for another type of work at the same location. This is one reason why it is helpful to note all of the possibilities from the outset. Numerous factors will help determine what is the most appropriate or relevant means of fall protection in any given situation.

19-5 SUMMARY

The fall hazard survey, including means and methods of protection, should be completed prior to work commencing, and should be thoroughly reviewed with any

SUMMARY

Fall Hazard Mitigation

Hazard ID _____

Prepared by: _____   Date_____

1. **Elimination or Substitution.** Can the hazard or hazardous work practices be removed or can a substitute work method be used?

   < > Yes Comments: _____
   < > No   Explain why: _____

2. **Passive Fall Protection.** Can the the hazard or hazardous work practice be isolated or separated from employees or others?(guardrail, cover, etc.)

   < > Yes Comments: _____
   < > No   Explain why: _____

3. **Positioning.** Can this Fall Hazard be mitigated through the use of a Positioning System? (max free fall 2 ft)

   < > Yes Comments: _____
   < > No   Explain why: _____

4. **Restraint.** Can this Fall Hazard be mitigated through the use of Restraint Systems? (Max slope 4:12)

   < > Yes Comments: _____
   < > No   Explain why: _____

5. **Rope Access.** Can this Fall Hazard be mitigated through the use of a Rope Access System?

   < > Yes Comments: _____
   < > No   Explain why: _____

6. **Fall Arrest.** Can this Fall Hazard be mitigated through the use of a Fall Arrest System?

   < > Yes Comments: _____
   < > No   Explain why: _____

NOTE: It is recommended that work that cannot be protected by one of the above methods be avoided at all costs.

FIGURE 19-3 Fall Hazard mitigation.

employee(s) who are likely to be exposed to the hazard so that they are aware of the hazards. Before any exposure is permitted a mitigation plan should be developed. This plan may be generalized, and apply to all work at that location, or it may be specific to the type of work. The survey information should also be made available to visitors who may be exposed, whether they are employees from other facilities, subcontractors, or others.

Subcontractors and others who are working on a site should always perform their own fall hazard survey, including identification of mitigation and protection methods, but it is good to consider the survey from the host organization where it is available. Visitors to a worksite should take the initiative to mitigate and protect themselves from potential exposure before commencing work.

CHAPTER 20

Creating a Rescue Preplan

Developing a rescue plan is an integral part of the managed fall protection program. Although rope access technicians are typically well versed in coworker-assisted rescue methods, this does not relieve the employer of the obligation of developing and ensuring a complete plan for rescue. Of course, the plan may – and should – incorporate the rope rescue capabilities of technicians already on site, as appropriate, but it should also go further and consider what is to be done if additional resources are necessary, determine whether any additional or specialized equipment is required, and establish how medical and trauma care will be provided. These capabilities may come from either on-site or off-site resources.

While it is not mandated that a rescue preplan be *written*, a written plan is arguably the best way by which to monitor and develop procedures, ensure that appropriate equipment is available, and to provide consistent and adequate training to affected and involved personnel. Each authorized person working at height should receive training in accordance with his or her respective role(s) and responsibility(s) within a fall protection plan. Preparing to respond to a fall is clearly an important aspect of an adequate fall protection plan. For an employer to effectively meet the rescue requirements of the fall protection plan, some preparations must be made in advance.

20-1 EMERGENCY RESPONSE PLANNING

It is generally considered good practice for a site owner or manager to develop an Emergency Action Plan (EAP) to address foreseeable workplace emergencies. Exactly what is covered in a given EAP will vary from one workplace to another, depending on the needs of each workplace.

According to OSHA in the United States, an EAP is a written document that must include but is not limited to[1]:

---

[1] Occupational Safety and Health Administration, Department of Labor *1910.38 employee emergency plans in general industry 1926.20, .21, and .35 training and emergency action plans in construction*

Professional Rope Access: A Guide To Working Safely at Height, First Edition. Loui McCurley.
© 2016 John Wiley & Sons, Inc. Published 2016 by John Wiley & Sons, Inc.

1. Means of reporting fires and other emergencies
2. Escape procedures and personal escape route assignments
3. Procedures to be followed by employees who remain on-site to conduct critical operations before they evacuate
4. Rescue and medical duties for those employees who are to perform them
5. Procedures to account for all employees after emergency evacuation has been completed
6. Names and/or job titles of persons who can be contacted for further information or explanation of duties under the plan.

A template for creating an EAP may be found in Figure 20-1. Each potential incident identified should be addressed separately, and additional information added as appropriate.

The site owner or manager must ensure that potentially impacted individuals are trained to carry out the established emergency plan. The plan should be re-evaluated periodically to ensure that it aligns with the potential needs and that assigned employees are able to perform to the necessary levels in the event of a real emergency.

EMERGENCY ACTION PLAN

INCIDENT TYPE: ____________________

ALARM/NOTIFICATION METHOD: ____________________

DESIGNATED RESPONDER(S):

Name	Responsibility

EVACUATION PROTOCOL/MUSTER AREA (MAP AS NEEDED)

TRAINING REQUIRED:

FOR MORE INFORMATION, CONTACT:

NAME
ADDRESS
ADDRESS
TELEPHONE
EMAIL/WEB

FIGURE 20-1
Emergency Action Plan template.

Most EAPs simply address the most obvious and common potential incidents such as flood, fire, tornado, hurricane, etc. Although EAPs do not necessarily typically address post-fall rescue, a potential for falls deserves no less consideration. Where potential hazards involve a possible fall from height or confined space incident, EAP guidelines provide an excellent framework for planning.

20-2 FALL RESCUE PLANNING

Regulatory requirements, as well as ethical obligations, dictate that whenever workers are placed in a potentially hazardous situation, including work at height, provision must be made for prompt rescue, as discussed in Chapter 13.

29CFR 1910.66
App C Sect I (e) (8)
Occupational Safety and Health Standards
Subpart: F – Personal Fall Arrest Systems

The employer shall provide for prompt rescue of employees in the event of a fall or shall assure the self-rescue capability of employees.

1926.502(d)(20)
Safety and Health Regulations for Construction
Subpart: M – Fall Protection

The employer shall provide for prompt rescue of employees in the event of a fall or shall assure that employees are able to rescue themselves.

A post-fall rescue plan must be grounded in the golden rule of rescue; that is that *no further harm* will be created by the rescue attempt – either to the subject or the rescuers.

When it comes to post-fall rescue, it is difficult to foresee every possible scenario or incident, so setting forth specific guidelines can be challenging. For this reason, emphasis should be placed on the concept of providing a range of capabilities that can be customized and applied to different circumstances as needed. To this end, preparing for both self-rescue and coworker-assisted rescue is prudent. While self-rescue is certainly an admirable, preferred goal, an alternative plan must exist in the event that a worker is incapacitated by an injury, illness, or even shock.

With this in mind, the rescue preplan should outline a progressive approach to responding to a fall, beginning with the idea that ALL authorized at height workers should be capable of personal escape – that is, escaping from his predicament in the event of a fall – AND that in most cases it is advisable to ALSO have a plan for immediate, on-site response, usually provided through coworker-assisted rescue. Finally, employers should make advance preparation for an expanded professional rescue capability, either through on-site standby rescue services or off-site resources such as local emergency services responders.

Rescue Pre-Plan
Overview

Local Municipal Resources

	Agency Name	Contact Person	Phone	How to Activate
Ambulance/Medical				
Nearest Hospital				
High Angle Rescue				

Identify what agency will be used for:

Self-Rescue	
Initial Response 5-45 minutes	
Professional Rescue 45+ minutes	
Expanded Capability	
First-Aid	

Complete the following for each hazard identified:

Hazard ID _____
(refer to Hazard Survey)

What immediate action(s) should a worker take to notify co-workers in the event of a fall or other emergency?

What immediate action should co-workers take to notify company representatives, emergency services, or other authorities in the event of a fall?

What is the minimum and maximum amount of time it will take for responders from municipal agencies to reach the suspended worker?
                Minimum   Maximum
High Angle Rescue:
Ambulance/Medical:

Where can EMS agencies expect to access MSDs sheets and other specific information applicable to the worksite?

FIGURE 20-2 Rescue preplan.

Wherever work at height is taking place, the corresponding rescue plan should include provision for

- personal escape
- Coworker-assisted rescue
- Professional rescue

Specific rescue methods are discussed in Chapter 13. The plan for achieving post-fall rescue should include, at a minimum, protocols to report the incident, personal escape techniques to be used by the fallen employee, assisted rescue techniques that can be implemented within just a few minutes, expanded technical rescue techniques for more complex scenarios, requirements for communication among group members and medical/first aid provision. A tool for outlining and developing such a plan is provided in Figure 20-2.

Additional information pertaining to planning for post-fall rescue may be found in the text Falls from Height: A Guide to Rescue (McCurley, 2013[2])

---

[2]McCurley, L. (ed) (2013) Index, in Falls from Height: A Guide to Rescue Planning, John Wiley & Sons, Inc., Hoboken, NJ, USA. doi: 10.1002/9781118640999.index

FALL RESCUE PLANNING

Complete the applicable section(s) below for the above incident

Describe in detail Self-Rescue Procedures for a suspended worker (diagram as appropriate):
Equipment:

Methods:

These methods should be practiced (circle one) Monthly/Semi-Annually/Annually by all affected workers

Location(s) of on-site Rescue Cache:

Describe Co-Worker Rescue Procedures for a suspended worker from above using raising techniques (diagram as appropriate):
Equipment:

Methods:

These methods should be practiced (circle one) Monthly/Semi-Annually/Annually by all affected workers

Describe Co-Worker Rescue Procedures for a suspended worker from above using lowering techniques (diagram as appropriate):
Equipment:

Methods:

These methods should be practiced (circle one) Monthly/Semi-Annually/Annually by all affected workers

FIGURE 20-2
(Continued)

Developing a Fall Rescue Response Plan

1. Define area of responsibility
2. Identify known or foreseeable fall hazards. Consider
   (a) # persons exposed
   (b) contributory factors (environment, task, experience, etc.)
   (c) time to rescue
3. For each hazard, develop a PERSONAL ESCAPE protocol, an ASSISTED RESCUE protocol, and an EXTERNAL RESPONSE protocol, including methodologies at every level for:
   - Notification of incident
   - Individual responsibilities
   - Equipment and systems to be used
   - Criteria and means for activation of the "next level" of response
4. Train, practice, assess, re-train

It is not enough to simply have the right equipment on hand, nor is it sufficient to have participated in a class or two on the subject. True *rescue capability* requires a coordinated combination of the right equipment and the proficiency to use it.

FIGURE 20-2
(Continued)

> Describe Co-Worker Rescue Procedures for a suspended worker from above using descent techniques (diagram as appropriate):
> Equipment:
>
> Methods:
>
> *These methods should be practiced (circle one) Monthly/Semi-Annually/Annually by all affected workers*
>
> Describe Co-Worker Rescue Procedures for a suspended worker from above using lowering techniques (diagram as appropriate):
> Equipment:
>
> Methods:
>
> *These methods should be practiced (circle one) Monthly / Semi-Annually / Annually by all affected workers*
>
> Describe other approved, preplanned Co-Worker Rescue Procedures for a suspended Woker (diagram as appropriate):
> Equipment:
>
> Methods:
>
> *These methods should be practiced (circle one) Monthly/Semi-Annually/Annually by all affected workers*
>
> Describe in detail Professional Rescue Procedures for a suspended worker (diagram as appropriate):
> Equipment:
>
> Methods:
>
> *These methods should be practiced (circle one) Monthly/Semi-Annually/Annually by all affected workers*

A training program for Coworker-Assisted Rescue responders should include guidance and hands-on practice for:

1. sizing up existing and potential conditions at an incident, including scope and magnitude of the incident, information about the physical and mental condition of the subject, what resources will be required, how long the operation can be expected to take, and environmental factors;
2. initial and ongoing hazard assessment;
3. methods for effecting the rescue without imposing undue hazards to rescuers or bystanders;
4. activating the next level of emergency response;
5. demobilization and debrief of involved resources;
6. documentation.

Each person who may be called upon to perform Coworker-assisted rescue should be afforded an opportunity to practice performing simulated rescues at least once every 12 months. In high-risk environments, practice should be even more frequent. Practice should involve removing manikins or actual persons from environments and structures that closely resemble those from which he might be required to perform an assisted rescue in an actual emergency.

20-3 COORDINATION WITH EXTERNAL RESOURCES

Although activation of local emergency services is not, in and of itself, considered to be a complete "Rescue Plan," advance coordination with local emergency services providers is an important part of any rescue plan. Site owners/managers should contact local emergency services organizations in advance to discuss and ensure a timely response and proper preparedness upon arrival, especially in workplaces where there is high probability of a worker being exposed to a potential fall.

Not all municipal response agencies have ready access to high angle rope rescue capabilities, and if response to this type of incident requires a separate call-out it would be best to indicate that from the outset rather than waiting until the first responders get there and then having to send out a second alarm. It is also worth putting reasonable effort into preparing local responders for what they will encounter when they arrive. Allowing response organizations to conduct annual training on-site, and providing tours for them anytime exposures change, can significantly reduce the response timeline in the event of a true high angle emergency.

The capabilities of rescue services vary widely between jurisdictions, and not all agencies are capable of responding to all types of incidents. Municipal response organizations should at the very least be informed of the hazards that exist in your workplace, and if possible dialog and/or even cooperative practice sessions should periodically take place to ensure adequate preparation on all fronts.

As a part of this process, a meeting with local area rescue/emergency service providers should be held to share information, set forth expectations, and to plan and coordinate the required evaluations. At this meeting, the employer should discuss key aspects of rescue with the organization, understanding that the goal is to gain understanding rather than to assume or insist that a certain capability be mandated.

In preparation for this discussion, the employer should understand his own needs with respect to the amount of time it may take for the rescue service to receive notification, arrive at the scene, set up, and be ready for rescue. The employer should be prepared to make clear to the response organization how to get from the entryway of the facility to the location or locations where rescue might most likely be necessary.

Likewise, the response organization should provide honest and straightforward information about their capabilities and limitations, including response time to the employer's workplace with respect to distance, roadways, and traffic. It is quite possible that the rescue service may predict different response times on different days of the week or at different hours of the day, or if they are already engaged in another response at the time of a call.

Finally, there should be open discussion of whether the rescue service does or does not own the equipment necessary to perform rescues, or if the equipment must be provided by the employer or another source.

The format provided in Figure 20-3 helps to ensure that key points are addressed when consulting with potential rescue resources. Complete a separate worksheet with each Professional Rescue Agency potentially involved.

20-4 SUMMARY

Effective post-fall rescue capability is a critical part of any effective managed fall protection program. Post-fall rescue capability is not something that creates itself.

Pre-Plan Worksheet for Liason With Professional Rescue Agency

Use this worksheet to ensure that key points are addressed when consulting with potential rescue resources
Complete a separate sheet with each Professional Rescue Agency potentially involved

Name of Agency	
Contact Person/Liason	
Office Phone	
Agency Activation Method	
Date of Meeting	
Worksite Address	
Covered Dates	

Meeting Attendees

    Program Administrator:_____

    Agency Liason:_____

    Auditor:_____

The following questions should be reviewed, mutually agreed, and signed off by all attendees.

	Administrator	Liason	Auditor
1. Agency is familiar with worksite and has reviewed Fall Hazard Survey			
2. Agency is capable of responding to all potential falls covered by the Fall Hazard Survey, with the following exceptions: (if none, write "none". Use addendum if needed.)			
3. Maximum response time to areas covered by the Managed Fall Protection Plan is _____ minutes.			
4. Agency is familiar with all fall protection equipment used on worksite.			
5. Agency must be notified _____hrs in advance when the following types of activities are planned:			
6. Agency will use the following methods to rescue suspended workers (use addendum if needed):			
7. Agency will maintain capability for this type of rescue by practicing on at least a (circle one) weekly / monthly / quarterly / annual basis.			
NOTES:			

Signed:

    Program Administrator:_____Date:_____

    Agency Liason:_____Date:_____

    Auditor:_____ Date:_____

FIGURE 20-3
Worksheet for coordinating with external agencies.

It requires intentional planning and preparation, requiring a combination of technical rope rescue ability and at least basic medical capabilities. While neither of these necessarily need be provided at advanced levels from internal resources, consideration should be given to both. A highly trained rigger or vertical technician is just that and should not be considered a "rescuer" unless they also possess specific rescue and medical training to make them adept at patient evaluation, packaging, and care. Likewise, a paramedic or other medical professional is just that, and should not be considered a "rescuer" unless they also possess specific rigging and rescue training to make them adept at extrication, evacuation, lowering, and raising techniques.

CHAPTER 21

Training Records

Technician competence requires more than just a bit of training and certification as a rope access technician. True competence requires a combination of knowledge, skills, and experience. These may be achieved in any number of ways, and any technician who can pass a rope access certification test without specific training should be commended. Even so, the importance of training should not be underestimated.

To prevent incidents, it is imperative that all users of rope access for height safety be required to be properly educated, equipped, certified, and maintain continual education/training in their use of rope-based systems. Training is best verified through independent testing/verification of skills rather than be measured in hours.

In the world of litigation it is said that "if it isn't written down, it didn't happen." With this in mind, both the technician and the employer should strive to maintain appropriate records that reveal the extent and levels of knowledge, skill, and experience that workers possess.

21-1 CERTIFICATION RECORDS

International rope access standards call for independent evaluation and certification of technicians. The term "independent" may be manipulated and interpreted in any number of ways, but the intent of this statement is to say that the entity conferring the certification upon the technician should not have any commercial investment in the candidate technician, the candidate technician's company, or the organization providing the training. Absence of such conflict of interest helps to maintain a high level of confidence and quality assurance in the certification systems and in the capability of the certified technician.

There are a number of certification bodies throughout the world, most of whom adhere to international rope access certification guidelines (ISO, 2012-current). Certification received through an entity other than a professional rope access certification body, or any organization that does not follow the internationally accepted ISO 22486, should be considered with some reserve. A list of several known compliant rope access certification organizations is found in Figure 21-1.

Upon successful completion of the required evaluation and skills test, technicians are awarded a certification card (Figure 21-2) showing at what level they are

FIGURE 21-1
Rope access certification bodies.

- SPRAT - USA
- ARAA - Australia
- IRATA - United Kingdom
- SARA – South Africa
- FISAT - Germany
- SRAA - Singapore
- HKRAA – Hong Kong

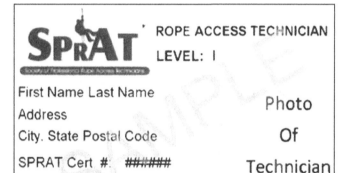

FIGURE 21-2
Sample certification card. Courtesy of SPRAT.

certified and the expiration date of said certification. These cards typically contain information about how to contact the certifying body if necessary. Certification bodies also maintain records of certified technicians so that an employer or prospective employer can verify the accuracy of certification claims if needed.

21-2 TECHNICIAN RECORDS

It is highly recommended that technicians maintain a recordbook containing thorough and accurate documentation of relevant training and experience. Recordbooks should contain information about each job, including type, duration, and location of the work, as well as the name and contact information of supervisors/employers, so that references may be checked by future prospective employers (Figure 21-3). Documented experience is very helpful in establishing the credibility and capability of personnel relative to future jobs.

Due to the migrant nature of rope access workers it is considered normal practice for technicians to retain their logbook in their possession, simply requesting supervisory and/or administrative staff at each jobsite to insert the requisite information. It is also normal for prospective employers, as well as certification organizations preparing to bestow a higher level of certification on that technician, to be permitted to review the logbook on request.

Record of Rope Access Work Experience					
Date	Company or Organization	Details of Work Tasks	Location	Hours Worked	Supervisor Signature
		Running total of hours worked			

FIGURE 21-3
Sample technician logbook. Courtesy of SPRAT.

21-3 EMPLOYER RECORDS

It is not just the technician who should keep records. Employers should document and maintain appropriate records for each technician within their employment, including (but not limited to):

1. statutory requirements relating to employment, next of kin, etc.;
2. training and competence records; and
3. experience.

Employers should document and maintain records of each employee's level of ability. This may include copies of relevant certification documentation, training records, work records, and/or results of employer-specific skills assessments. Training records should include, at a minimum, details regarding training dates, course content, and student performance. Figure 21-4 shows an example of a training record form.

New employers should review the technician's logbook and follow up with previous employers prior to inducting a technician into the field. Additional training should be provided where necessary, including when a technician is asked to perform a new skill or work in an unfamiliar environment.

Technicians with whom the employer is unfamiliar, as well as inexperienced technicians, should be continuously monitored until the rope access supervisor is satisfied with the level of competence of the technician. Newly certified technicians should initially work under the direct supervision of the rope access supervisor. Over time and with experience, the rope access supervisor may opt to permit the new technician to work under the close supervision of other experienced operatives.

Regardless of the level of skill or experience of personnel, technicians should be encouraged to monitor each other's rigging and safety systems on an ongoing basis to verify that they are rigged and used appropriately.

Figure 21-5 outlines the minimum training requirements for rope access technicians at three defined levels of capability (ISO, 2012-current). This list should be customized based on employer need, work environment, and circumstances.

```
TRAINING RECORD                    Employee ID:

Training Date: ___/___/_____      Duration of Training: _____ HRS

Student name:_____
Training Organization:_____ Trainer:_____

Course Objective(s)
[                                                              ]

Evaluating Organization:_____ Evaluator:_____
Evaluation by: (Check all that apply)

    WRITTEN TEST            PASS  (Circle One)  FAIL   (Attach copy if applicable)
    ORAL TEST               PASS  (Circle One)  FAIL   (Attach copy if applicable)
    SKILLS EVALUATION       PASS  (Circle One)  FAIL   (Attach copy if applicable)
    THEORETICAL EXERCISE    PASS  (Circle One)  FAIL   (Attach copy if applicable)

        [ EVALUATOR RECOMMENDATION    PASS    (Circle One)    FAIL ]

Signed _____ (Evaluator)   Date:_____

Recommended Frequency of Refresher Training:
        Rope Access Skills              _____ Months
        Fall Protection Methods         _____ Months
        Equipment Inspection Methods    _____ Months
        Rescue Methods                  _____ Months
        Administrative skills           _____ Months
        Other_____            _____ Months
```

FIGURE 21-4 Training record form.

21-4 PROGRAM ADMINISTRATOR TRAINING

The employer should designate a Program Administrator to be responsible for outlining and documenting all phases of the rope access program. At a minimum, Program Administrators should have a working knowledge of current fall protection regulations and standards, rope access systems and processes, as well as specific knowledge applicable to the industry and the job in which they are performing.

Ideally, a Program Administrator should have been trained and certified in rope access at some point in his or her career, but it is not imperative that the Program Administrator be a field active technician. Management skills are the more essential skills at this level. With this in mind, the Program Administrator will create and oversee the rope access program in its entirety, including essential records, planning documentation, personnel files, risk assessments, and related information.

	Basic	Intermediate	Advanced
Methods for Work Restraint	C	C	C
Methods for Work Positioning	C	C	C
Methods for Fall Arrest	C	C	C
Methods for Rope Access Backup	C	C	C
Equipment Donning/Doffing	C	C	C
Pre-Use Inspection	C	C	C
Safe work within Access Zone	C	C	C
Methods of Descent	C	C	C
Methods of Ascent	C	C	C
Changeovers	C	C	C
Knot Pass	C	C	C
Anchorage selection	C	C	C
Elementary Rigging	C	C	C
Rope Management	C	C	C
Pass a ReAnchor	C	C	C
Place and Pass a Deviation	C	C	C
Install a ReAnchor	K	C	C
Simple Mechanical Advantage	K	C	C
Rope to Rope Transfer	C	C	C
Vertical Rescue from Descent	C	C	C
Vertical Rescue from Ascent	K	C	C
Rescue through Obstruction	K	C	C
Rescue remote casualty by lifting	K	C	C
Rescue through horizontal and vertical	K	K	C
Team rescue management	NA	K	C
Team rescue participation K C C	K	C	C
Aid climbing	K	C	C
Horizontal Traverse	K	C	C
Lead climbing	NA	K	C
Advanced Rigging	NA	C	C
Highlines	NA	C	C

Key NA= Not Applicable
K = Knowledge (Awareness)
C = Competence (Skill)

FIGURE 21-5
Minimum training guidelines for rope access technicians at different levels (ISO 22486).

The employer should specify qualifications and training for the Program Administrator, including specific criteria for knowledge and skill related to rope access, conventional fall protection, rescue, equipment, and systems applicable to fall hazards experienced in the job. These criteria should be based on the types of fall protection and other hazards that they are likely to experience in that position.

At a minimum the Program Administrator should possess the following knowledge and skills:

- Familiar with all skills expected of an authorized rope access technician
- Use and administration of the Managed Fall Protection/Rope Access program
- Use and maintenance of company's Fall Hazard Survey(s)
- Selection and appointment of safety committee personnel
- Rope access system selection in accordance with employer's program
- Development of approved equipment purchase lists
- Selection and appointment of authorized, competent, and qualified persons
- Administration of rope access plans and procedures outlined by this document
- Administration of employer's training programs.

CHAPTER 22
Equipment Inspection and Care

Chapters 5 and 6 discuss many of the finer details of selecting equipment for rope access, but this serves only to narrow the range of acceptable choices. With so many options to choose from, how does one choose which components to accept on the job-site, and what are the best practices for keeping that equipment serviceable and safe?

This chapter will provide practicable tools for specifying the equipment, putting the equipment into service, and for inspecting and maintaining it on an ongoing basis.

22-1 SPECIFYING EQUIPMENT

All items of equipment that are used to support a person (i.e., ropes, harnesses, descenders, and other attachments) should be strong enough to provide an adequate safety factor over the most severe combination of loads that is reasonable to predict. If there is any question as to the applicability of reported performance information relative to the intended use, the manufacturer should be consulted for additional information.

Consideration must also be given to compliance with applicable regulations and standards in the location where work is being performed. Note that there are not necessarily applicable regulatory requirements for all components of equipment in all jurisdictions, but it is the responsibility of the Program Manager to research and verify this as necessary. Keep in mind that it is the location where work is being performed that matters, which is not necessarily the same jurisdiction as the headquarters of the company performing the work. Working outside one's own jurisdiction can require extensive research in this area.

If technicians are permitted to bring their own equipment to the job, the Program Manager should ensure its adequacy before it is used. It may be necessary for the Program Manager to consult with equipment manufacturers or to study up on different devices before approving or disapproving their use for a given project. Devices should be marked with the manufacturer's identification as well as a unique identifier to allow traceability to their test, inspection, or certificates of conformity.

Professional Rope Access: A Guide To Working Safely at Height, First Edition. Loui McCurley.
© 2016 John Wiley & Sons, Inc. Published 2016 by John Wiley & Sons, Inc.

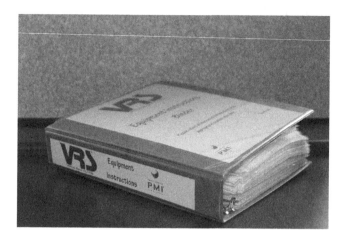

FIGURE 22-1
Maintain copies of all product instructions.

All technicians on a given site should be given ample opportunity to familiarize themselves with equipment that other technicians are using. This will help to ensure adequate "buddy checks" during work, as well as facilitate rescue in the event of an emergency. It is good practice to maintain a collection of product instructions for each unique component of the worksite, in a location where they can be referenced by technicians as needed. This may be done with hard copies, as shown in Figure 22-1, or electronically.

22-2 PLACING EQUIPMENT INTO SERVICE

Prior to being placed into service, equipment should be marked with some means of field traceability so that it can be monitored over the course of its lifespan. This may be achieved by etching or engraving or tape or permanent marker or some other method, as long as it is sufficiently durable to last the life of the product or can be easily refreshed.

The date that the equipment is placed into service should be recorded, and instructions for that piece of equipment should be retained and stored where they may be referenced in the future. One way to maintain such records is to use an Equipment Information Log for each component. One example of such a log is shown in Figure 22-2.

While the example shown above is specific to rope, similar tools can be developed and used for other components of equipment. There should be a place on the log to note at least the name of the product and the manufacturer, date purchased, date in service, unique identifier, and other relevant information. It is also useful to use the same form to record location, nature, and duration of use, as well as details of periodic inspections. These records can then be maintained either in a notebook or in an electronic format. The goal in maintaining the records should be to give technicians, supervisors, safety managers, and employers ready access to relevant information on every item used on the jobsite.

22-3 EQUIPMENT INSPECTION

Rope access equipment should be inspected before each use by a competent user. In fact, most competent rope access technicians will insist on inspecting any piece

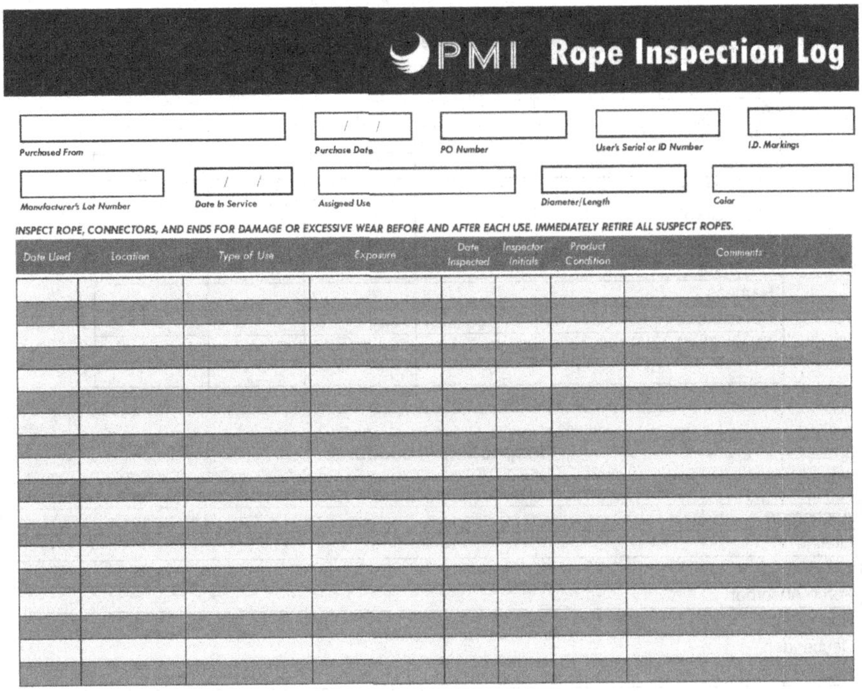

PMI Rope Inspection Log

Purchased From | Purchase Date / / | PO Number | User's Serial or ID Number | I.D. Markings

Manufacturer's Lot Number | Date In Service / / | Assigned Use | Diameter/Length | Color

INSPECT ROPE, CONNECTORS, AND ENDS FOR DAMAGE OR EXCESSIVE WEAR BEFORE AND AFTER EACH USE. IMMEDIATELY RETIRE ALL SUSPECT ROPES.

Date Used	Location	Type of Use	Exposure	Date Inspected	Inspector Initials	Product Condition	Comments

FIGURE 22-2
Sample Equipment Information Log.

of equipment upon which their life will depend. It is also not uncommon for a rope access supervisor to perform a preuse check on equipment that will be rigged on a site for which they are responsible. Such preuse inspections may simply involve a cursory once-over and an abbreviated function check – just enough to ensure that the equipment is in good working order and ready for use. Preuse inspections do not necessarily require formal documentation, although use of a checklist can be advantageous to ensure thorough evaluation. Any item that is found to not be in good working condition should not be used.

In addition to preuse checks, all equipment must undergo a more thorough, periodic inspection by a competent person on a regular basis, as defined by the employer in accordance with manufacturer's instructions. Periodic inspection is a more detailed evaluation that should include a review of purchase and in-service dates, close examination of both form and function of the device, evaluation of wear and tear, and other relevant information. Normally, inspection intervals of 6 months are considered "reasonable," but these should be performed certainly no less than annually. Inspection results should be recorded in a manner consistent with the employer's recordkeeping methods, such as that illustrated in Figure 22-2.

Inspection procedures and retirement processes should be outlined for each component, including how often each component should be inspected, what to look for, and how the information should be recorded. This may be set out in the employers Comprehensive Managed Fall Protection Program using Equipment Inspection Guidance Worksheets such as those shown in Figure 22-3.

22-4 CLEANING

Dirt and grime can impede proper function of equipment and can cause it to wear prematurely. Rope access equipment may be cleaned as needed, as long as guidance from the manufacturer is followed. While dirt and grime is relatively easy to

Recommended Inspection Intervals
(for each item check all that apply)

	Before each use	6 month periodic	Annual Periodic	Other
Anchorage				
Lifeline				
Lanyard				
Force Absorber				
Rope Grab				
Descender				
Ascender				
Harness				
Connectors				
Other: (list)				

Inspection Procedures

	Guidance (Consider visual and tactile methods. Include manufacturers recommendations and local protocols)
Anchorage	
Lifeline	
Lanyard	
Force Absorber	
Rope Grab	
Descender	
Ascender	
Harness	
Connectors	
Other: (list)	

FIGURE 22-3
Equipment Inspection Guidance Worksheets.

remove from equipment, serious contaminants such as chemicals can be more difficult to eliminate completely. If equipment is exposed to contaminants that are likely to damage it through long-term exposure, the conservative approach is to simply retire and replace it.

Equipment made entirely from metal may be submerged for a few minutes in hot soapy water, then thoroughly rinsed and air-dried before storage. High-pressure washers, steam cleaners, seawater, and abrasives should generally not be used unless specifically recommended by the manufacturer. Textiles (rope, harness, webbing, etc.,) may be washed in lukewarm water with mild, nondetergent soap, then thoroughly rinsed in clean water and air-dried away from direct heat.

Equipment should not be stored in a wet condition.

22-5 RETIREMENT

The employer should establish retirement procedures to ensure that retired equipment is not inadvertently used or placed back into service. Inspection and retirement should, at a minimum, be in compliance with manufacturer's instructions. The employer may set out additional criteria based on known workplace hazards or other conditions.

Equipment that does not pass inspection should be removed from service. Alterations and repairs should not be made to any equipment by the end user or without prior approval of the manufacturer. In some cases, manufacturers are able to repair or refurbish the equipment. In such cases, the equipment should be inspected by a competent person prior to being placed back into service.

For some equipment, manufacturers may have established a maximum lifespan or obsolescence date. Equipment should not be retained or used beyond such dates, even if it continues to pass tactile and visual inspection.

It is often advisable to destroy or render inoperable retired equipment to prevent subsequent use. For example, cutting life safety rope into short lengths that would be impractical for use, or removing essential parts of hardware, can help ensure that an uninformed person does not pick it up and inadvertently put it into use.

CHAPTER 23

Rope Access Program Audit

The development of a Rope Access Program within the context of an employer's Comprehensive Managed Fall Protection Program is an important step toward improving safety in work at height. Periodically auditing the rope access program will help to ensure that the program remains up to date and that it is being effectively implemented.

An audit may be performed by the Program Administrator or, for a more objective perspective, by an external resource.

An audit should include a full review of all applicable procedures, complemented by some means of verifying that employees are actually adhering to the process during regular working days. Recommended audit intervals are at least once every two years, whenever Authorized Persons or Competent Persons identify deficiencies, or when there are incidents involving injuries, property damage, or "near misses."

Criteria for an audit should be based on a reasonable set of external, consensus requirements, such as the International Rope Access Standards (ISO, 2012-current), and the goal for the audit should be one of encouragement toward excellence rather than being punitive in nature. Audit criteria should be well defined and applied in an objective manner.

23-1 COMPONENTS OF AN AUDIT

An audit should contain a document review, observation of rope access work, management interviews, and employee interviews. Using these methods, the auditor will seek to affirm that a company has established appropriate processes and procedures for their rope access program, that they are adhering to these processes and procedures, and that relevant information is being communicated effectively to stakeholders throughout the organization.

Topics typically covered by an audit will include management, practices, equipment, and recordkeeping. An example of an audit form is be found in Figure 23-1.

The remainder of this chapter will discuss each segment of the audit, and how to perform an effective audit.

Professional Rope Access: A Guide To Working Safely at Height, First Edition. Loui McCurley.
© 2016 John Wiley & Sons, Inc. Published 2016 by John Wiley & Sons, Inc.

	Rope Access Company Audit		

Organization/Company: _____
Representative: _____ Title_____
Location/Address of Audit:_____
        Audit Dates   Start____/____/_____   Complete____/____/_____
Auditor: _____ Representing:_____
Auditor Contact Information:_____

	Does not meet Expectations	Meets Expectations	Exceeds Expectations
Program Management			
Organizational Structure			
Comprehensive Managed Fall Protection Plan			
Policy Statement			
Job Hazard Analysis			
Fall Hazard Survey			
Incident/Near Miss Investigation			
SCORE			
Rope Access Practices			
Training Requirements			
Leadership & Supervision			
Rope Access Work Plan (Permit)			
Technician Skills			
Rescue Planning			
SCORE			
Equipment Management			
Personal Equipment			
Rigging Equipment			
Storage & Maintenance			
Inspection Procedures			
Retirement Criteria/Processes			
SCORE			
Recordkeeping			
Technician Records			
Project Records			
Training Records			
Equipment Inspection Records			
SCORE			

NOTES

AUDITOR RECOMMENDATION: PASS FAIL   Signed:_____

FIGURE 23-1 Sample audit form.

23-2 MANAGEMENT

The first step in conducting an audit should be to review management structure and practices, as this forms the basis for the entire rope access program.

- Organizational Structure
  The auditor should verify that the organizational structure of the rope access program clearly identifies the responsibilities for at least safety management, program administration, and supervision of rope access. Such verification may be established by means of an organizational chart or other records. These responsibilities may be held by one person or they may be divided between individuals, but in any case the lines of authority should be made clear. The auditor may interview various staff members to ensure thorough understanding and awareness of the organizational structure.

- Comprehensive Managed Fall Protection Plan
  The organization should be able to provide written documentation comprising their Comprehensive Managed Fall Protection Plan. The content of this plan will consist primarily of specific documentation otherwise called out in the audit sheet, but the idea here is that the organization will have compiled a documentation that summarizes how work at height is managed, the roles and responsibilities, safety considerations, and what fall protection means and measures are considered acceptable within the organization.
- Policy Statement
  As part of the Comprehensive Managed Fall Protection Plan, a document should exist that establishes the position of rope access relative to the company's fall protection program, and states general goals of the rope access program as well as the commitment of the management to safety. A copy of this document should be on file, and employees should be able to summarize the intent of this document on request.
- Job Hazard Analysis
  The auditor may request that the organization present an identifiable process and tools for performing a JHA. In many cases, a general purpose JHA will be included as part of the Comprehensive Managed Fall Protection Program, and additional JHAs will be completed relative to individual projects. Personnel at all levels of the organization should be able to explain to the auditor the relevance of the JHA, commensurate with their level of involvement. For example, the Program Manager should be able to explain the purpose and need for the document, the Supervisor should be able to describe how to perform the analysis, and Technicians should discuss how they use the document. An example of a JHA from an actual project should be available on request, and the auditor should verify that this document is completed appropriately.
- Fall Hazard Survey
  Similar to the JHA, the Fall Hazard Survey is a document with which affected personnel should be familiar. Many organizations will maintain a corporate fall hazard survey for general use, but will perform individual fall hazard surveys for specific projects. The auditor should review the overall fall hazard survey, as well as a fall hazard survey document from a previous project, to affirm appropriate use of the tool. Personnel at various levels should be interviewed to show that they understand the need for the document and how it is used.
- Incident/Near Miss Investigation
  The organization must have a process for recording and investigating incidents and near misses, with the goal of preventing future incidents. The auditor will verify that the process used is appropriate and thorough, and will review the list of near misses incidents/accidents and corrective action reports. At least one example of the records from an actual investigation (if possible) will be selected and the documentation of that incident/accident will be reviewed. The process and documentation should be consistent with the organization's stated incident investigation process and should include a plan for preventing similar incidents in the future.

23-3 PRACTICES

The next phase of an audit involves consideration of specific rope access practices to ensure that they are in alignment with accepted principles and industry best practice. Affirming practices may require a review of past projects, interviews with personnel, and/or observation of work.

- Training Requirements
  The organization should exhibit a commitment to best practice, including requirements for certification of technicians and a commitment to providing ongoing refresher information, training, and updates on an ongoing basis. The auditor should be able to verify that personnel who are designated as "rope access technicians" are appropriately certified, and that technicians are used within the scope of their respective levels of training.

- Leadership and Supervision
  The Rope Access Program Administrator is ultimately responsible for the rope access program. This individual should be thoroughly interviewed to ascertain how they go about monitoring the performance of teams in the field, manage safety policies, and nurture staff development. Appropriate leadership and supervision include clear designation of responsibilities and authority, which the auditor should be able to identify through observation and interview with technicians and management. Oral findings should match written documentation. The practices found within the organization should reflect the recommended practice of using more experienced personnel to supervise less experienced personnel. Those designated as Authorized Persons, Competent Persons, and/or Qualified Persons should be able to identify themselves and one another, and the meanings of those terms.

- Rope Access Work Plan (Permit)
  The Rope Access Work Plan is usually specific to a given project; however, some organizations may maintain an over-arching plan for general use, or as a guide. The auditor should review the over-arching plan, if available, as well as an example of one or more specific project work plan(s), to establish whether work plans are used effectively and appropriately. For selected projects, the auditor will verify that all technicians performing rope access had current SPRAT certifications at the time of the job, the technicians possessed additional job-specific training if necessary for the work, appropriate management practices were used, JHA and Fall Hazard Surveys were used, and that an appropriate number of technicians were assigned to the project(s). Supervisory staff and technician staff involved in the particular project should also be interviewed to verify that the plan was appropriate to the work, and that the work followed the plan.

- Technician Skills
  Certification is an excellent tool for ensuring that personnel are knowledgeable and skilled at a certain level of performance. However, the auditor may also choose to verify technician capability by requesting that selected technicians perform certain skills corresponding to their level of certification. Observation of skills demonstrations can give the auditor greater insight into the depth of ability and competence within an organization.

- Rescue Planning
  Rope access technician certification generally includes a requirement that the candidate demonstrates the ability to rescue a coworker. However, the need for rescue goes beyond simple rope skills, and should include provision for emergency evacuation and medical care in the event that they becomes necessary. The auditor should review a sample rescue plan from a previous project

to verify that the organization uses rescue plans to include coworker-assisted rescue as well as contingency planning for municipal response, emergency medical response, and evacuation to advanced care if needed.

23-4 EQUIPMENT MANAGEMENT

Overall responsibility for equipment lies with the Program Manager. However, this duty may be delegated to other personnel as appropriate. Regardless of who performs the actions, the organization should be able to demonstrate good practice when it comes to managing equipment.

- Personal Equipment
  In some cases, personal equipment is provided by the employer while in other cases personal equipment may be provided by the technicians themselves. Either way, it is the employer's responsibility to ensure that the equipment used is appropriate and acceptable. The auditor should verify that practices exist to ensure appropriate selection of equipment for use by technicians. There should also be established methods for ensuring that technicians perform a preuse check of all personal equipment, as well as submit a verification to the employer that personal equipment is thoroughly inspected on a periodic basis by a Competent Person. Finally, the auditor should observe and confirm that equipment is used properly.
- Rigging Equipment
  Rigging equipment is generally provided by the employer, who must also take measures to ensure that technicians who will use the equipment are familiar with its use and limitations. Technicians should exhibit a pattern of inspecting rigging equipment before use. The organization should be able to provide evidence of periodic inspections by a Competent Person.
- Storage and Maintenance
  The auditor should observe and ask questions regarding storage and maintenance of both personal equipment and rigging equipment to ensure that it is stored and maintained properly. Product instructions should be requested for various pieces of equipment to exhibit that these are on hand and are available for review as needed. Cleaning procedures should be reviewed and cross-referenced with manufacturer instructions to verify appropriateness.
- Inspection Procedures
  Equipment inspection records should be available on request for both personal and rigging equipment. All technicians should be able to demonstrate appropriate preuse checks, and Competent Persons should be able to demonstrate procedures for performing and documenting periodic inspection. The auditor may perform spot checks of any equipment to affirm that equipment is in good condition and ready for use.
- Retirement Criteria/Processes
  The auditor may wish to review manufacturer recommendations for retirement of equipment and compare them with organization practices to ensure that the two are compatible. It is generally appropriate for organizations to use manufacturer guidelines conservatively, and to be aggressive in their retirement practices. The auditor should verify that equipment identified for retirement is disposed of properly, in a manner that prevents reuse.

23-5 RECORDKEEPING

While it is possible to have a good rope access program without adequate documentation, all of the matters pertaining to the management and verification of the Comprehensive Managed Fall Protection Program will be much easier to prove if adequate records are maintained. Recordkeeping is an essential part of ensuring consistency in good practices. To this end, the auditor should verify that adequate records are available (electronic or paper) with content that is appropriate to the size and complexity of the organization.

- Technician Records
  For all technicians on record, whether employees or contracted workers, the auditor should be able to find appropriate records that show levels and dates of rope access technician certification, substantiation of additional qualifications or certifications, emergency contact information, dates of employment or contract (as applicable) and project assignments.
- Project Records
  Upon request, the organization should be able to provide the auditor with some details of ongoing projects that include a rope access component. These details should include at least information pertaining to the geographic location of the work (region), work description (inspection, cleaning, painting, etc.), and access methods (simple descent; aid climbing; work positioning, etc.). The auditor may request additional information for one or more projects, at which time the organization should be capable of providing a complete work plan (see above) for that project.
- Training Records
  The employer should not rely solely on technician certification, but should also maintain a verification procedure to review and monitor technician skills, and to provide additional training as appropriate. Such training may include skills refreshment, rigging procedures, company equipment familiarization, or other applicable topics. The auditor should review written policies and procedures regarding training, as well as the organization's training records. Sample records should be provided on request to show the relationship between employee's training/skills to their duties and responsibilities.

23-6 SUMMARY

This guide will help to prepare an organization for an effective audit to confirm the viability and performance of their rope access program. It may be used by an internal resource, such as a Program Manager or Safety Director, to perform an internal audit, or it may be used by an external resource to provide an independent audit. However, the information in this chapter is not specific to any one organization or audit process. Therefore, any company who is about to be audited by a specific organization should request audit policies and procedures from that organization to be adequately prepared for the audit.

Knowledge Check

KNOWLEDGE CHECK

Chapter 1 What Is Rope Access?

1. The term "rope access" is synonymous with the term(s)
   (a) Controlled descent
   (b) Bosun's chair
   (c) Rope descent systems (RDS)
   (d) None of the above
   (e) All of the above
2. Some distinct benefits offered by rope access include (check all that apply)
   (a) Reduced exposure time
   (b) Reduced cost
   (c) Higher levels of worker training
   (d) More workers on staff roll at any given time
   (e) Increased safety
   (f) Lower environmental impact
3. Rope access technicians are specifically trained to be capable of
   (a) Ascending, Descending, Self-Rescue, and Coworker-assisted Rescue
   (b) Leaping tall buildings in a single bound
   (c) Ascending, using dorsal fall arrest, and performing Self-Rescue
   (d) Ascending and Descending only
4. Professional rope access evolved from
   (a) Mountaineering and caving
   (b) Fall arrest and restraint techniques
   (c) Fall containment systems
   (d) None of the above
5. Modern rope access uses exactly the same equipment and techniques as
   (a) Recreational climbing
   (b) Recreational rappelling
   (c) Recreational caving
   (d) None of the above
6. The essential components of a rope access system include
   (a) A rope and interchangeable rope grabs
   (b) A fall arrest system and bosun's chair

Professional Rope Access: A Guide To Working Safely at Height, First Edition. Loui McCurley.
© 2016 John Wiley & Sons, Inc. Published 2016 by John Wiley & Sons, Inc.

(c) A primary access system and interchangeable secondary backup system

(d) None of the above

7. A rope access system often incorporates the use of a bosun's chair as the primary means of access

   (a) True
   (b) False

8. Rope access technician knowledge and skills should be verified

   (a) Through only a written test
   (b) By independent assessment and certification
   (c) Never
   (d) Only if the individual's ability has been called into question

9. Which of the following harness attachment points is most conducive to self-rescue

   (a) Dorsal attachment
   (b) Side D-rings
   (c) Sternal attachment
   (d) None of the above

10. Which type of backup safety system is used in a true rope access system?

    (a) Self-retracting lanyard
    (b) Interchangeable rope access backup system
    (c) Vertical lifeline fall arrest
    (d) Answers a and b are acceptable
    (e) Answers b and c are acceptable
    (f) Any/all of the above

Answer key:

1. d
2. a, b, c, e, f
3. a
4. a
5. d
6. c
7. b
8. b
9. c
10. b

KNOWLEDGE CHECK

Chapter 2 Rope Access and the Comprehensive Managed Fall Protection Plan

1. According to most regulatory requirements, the _____ is considered to be responsible for providing a safe place of employment
   (a) Owner
   (b) Employer
   (c) Manager
   (d) Employee
2. Where work at height is undertaken, Z359.2 recommends the development of a
   (a) Regulatory Contract for Fall Arrest
   (b) Memorandum of Understanding between site owners and managers
   (c) Comprehensive Managed Fall Protection Program
   (d) Program Management Guide
3. An "authorized person" is one who:
   (a) Is approved or assigned by the employer to perform a specific type of duty or duties or to be at a specific location or locations at the jobsite.
   (b) Is capable of identifying existing and predictable hazards in the surroundings or working conditions which are unsanitary, hazardous, or dangerous to employees, and who is authorized to take prompt corrective measures to eliminate them.
   (c) By possession of a recognized degree, certificate, or professional standing, and/or who by extensive knowledge, training and experience, has successfully demonstrated the ability to solve or resolve problems relating to the subject matter, the work, or the project.
   (d) None of the above
4. A "competent person" is one who:
   (a) Is approved or assigned by the employer to perform a specific type of duty or duties or to be at a specific location or locations at the jobsite.
   (b) Is capable of identifying existing and predictable hazards in the surroundings or working conditions, which are unsanitary, hazardous, or dangerous to employees, and who is authorized to take prompt corrective measures to eliminate them.
   (c) By possession of a recognized degree, certificate, or professional standing, and/or who by extensive knowledge, training and experience, has successfully demonstrated the ability to solve or resolve problems relating to the subject matter, the work, or the project.
   (d) None of the above
5. A "qualified person" is one who:
   (a) Is approved or assigned by the employer to perform a specific type of duty or duties or to be at a specific location or locations at the jobsite.
   (b) Is capable of identifying existing and predictable hazards in the surroundings or working conditions, which are unsanitary, hazardous, or dangerous to employees, and who is authorized to take prompt corrective measures to eliminate them.
   (c) by possession of a recognized degree, certificate, or professional standing, and/or who by extensive knowledge, training and experience, has

successfully demonstrated the ability to solve or resolve problems relating to the subject matter, the work, or the project.
   (d) None of the above
6. The duties of the Program Manager include
   (a) Developing the Comprehensive Managed Fall Protection Plan
   (b) Assigning tasks and responsibilities commensurate with the plan
   (c) Anchorage Selection and Rigging
   (d) Providing resources as needed to fulfill the requirements of the plan
   (e) a, b, c, d
   (f) a, b, d
   (g) a, c, d
7. A fall hazard survey should contain:
   (a) Pertinent information regarding types of fall hazards
   (b) Analysis of how severe the exposure potential might be
   (c) Recommended corrective solution(s)
   (d) All of the above
8. In the Hierarchy of protection against falls from height, active protection methods are
   (a) Not permitted
   (b) Considered the most desirable option
   (c) Considered the least desirable option
   (d) Not addressed
9. Harness-based methods for fall protection might include
   (a) Rope access, restraint, arial platform
   (b) Restraint, fall arrest, guardrails
   (c) Rope access, restraint, fall arrest
   (d) Prevention, passive, active
10. A Rope Access Work Plan should include
    (a) The rope access methods to be used, a list of work team members, and a job hazard analysis
    (b) A rope access equipment list, a list of previous work locations, and rescue information
    (c) A job hazard analysis, public safety information, and serial numbers of all PPE
    (d) A hierarchy of protection, public safety information, and a rescue plan

Answer key:

1. b
2. c
3. a
4. b
5. c
6. f
7. d
8. c
9. c
10. a

KNOWLEDGE CHECK

Chapter 3 Personnel Selection and Training

1. A prospective rope access technician must be able to
   (a) Read and understand product instructions
   (b) Understand and use job hazard worksheets
   (c) Comprehend a written rope access work plan
   (d) All of the above
2. The term "aptitude" refers to
   (a) Physical strength and agility
   (b) A measure of intelligence, exhibited through tests
   (c) Inherent suitability or inclination toward proficiency
   (d) Ability to engage in contract negotiations
3. Responsibilities of the rope access program manager include
   (a) Oversight of program, procedures, and work records
   (b) Purchasing of all PPE
   (c) Training and instruction for technicians
   (d) Evaluation and certification of technicians
4. A training program should include at least the following provisions
   (a) Training outline
   (b) Lunch
   (c) Performance requirements
   (d) a and c
   (e) b and c
5. The relationship between rope access and skilled trades is
   (a) Rope access is a skilled trade
   (b) Rope access is a method of accessing a location to apply a skilled trade
   (c) Skilled trades have no place in rope access
   (d) Rope access can only be used with skilled trades
6. A rope access team
   (a) Must always have at least two members and a designated supervisor
   (b) Must always have at least two members and a designated driver
   (c) Must always have at least four members and a designated supervisor
   (d) Must always have at least four members and a designated Program Manager
7. Technicians certified at the most basic level
   (a) Should always perform their own rigging
   (b) Should only work under the supervision of a more advanced technician
   (c) Should be responsible for writing the work plan
   (d) Should never attempt to ascend the rope
8. Technician logbook information should include at least
   (a) Emergency contact information, contact information for supervisors
   (b) Work experience, color-coded pages, contact information for supervisors
   (c) Training records, work experience, contact information for supervisors
   (d) Certification details, records, work experience

9. Technician training should be verified by
   (a) A specific number of hours required for training
   (b) Display of retained knowledge and demonstration of physical skills
   (c) A written test
   (d) A written attestation by a certified trainer
10. Certification of rope access technicians should be performed by an entity that is
    (a) Independent from the technician and/or training entity, and using ISO 22846 guidelines
    (b) Independent from the technician and/or training entity, and using ANSI Z359 guidelines
    (c) Owned by the training organization, and using ISO 22846 guidelines
    (d) Owned by the training organization, and using ANSI Z359 guidelines

Answer key:

1. d
2. c
3. a
4. d
5. b
6. a
7. b
8. c
9. b
10. a

KNOWLEDGE CHECK

Chapter 4 Equipment for Rigging

1. Regulatory requirements for rope access equipment
   (a) Do not exist
   (b) Exist, but need not be taken seriously in some jurisdictions
   (c) Are the same in all jurisdictions
   (d) May vary depending on the employer and regulatory authority in the place of work
2. Equipment used for life safety situations should be selected based on
   (a) Conformity with applicable regulatory requirements
   (b) Performance specifications
   (c) User familiarity and comfort level
   (d) All of the above
3. Equipment manufacturer recommendations should be followed regarding equipment
   (a) Selection
   (b) Use
   (c) Inspection and retirement
   (d) Maintenance
   (e) All of the above
4. Team Rigging Equipment items that are often shared among users include
   (a) Ropes, anchorage connectors, carabiners, and pulleys
   (b) Ropes, carabiners, and personal descenders
   (c) Ropes, descenders, and ascenders
   (d) Ropes, PPE, and anchorage connectors
5. Personal Equipment items that may be best assigned to individual users include
   (a) PPE, ropes, anchorage connectors, and descenders
   (b) Ropes, anchorage connectors, carabiners, and pulleys
   (c) PPE, descenders, ascenders, and carabiners
   (d) Anchorage connectors, carabiners, and descenders
6. Personal Protective Equipment (PPE)
   (a) Is just another term for all equipment owned and used by an individual person
   (b) Includes that Personal Equipment which is worn by the user for health and safety protection
   (c) Includes all equipment that is used on the worksite
   (d) Includes all items of equipment except that which is worn by the user
7. The best indicator of actual strength of an item is
   (a) Safe Working Load (SWL)
   (b) Working Load Limit (WLL)
   (c) Safety Factor (SF)
   (d) Minimum Breaking Strength (MBS)

8. Ropes used for primary access and backup during rope access are typically
   (a) Of laid construction and in the 10–12.5 mm range
   (b) Of laid construction and in the 8–11 mm range
   (c) Of kernmantle construction and in the 10–12.5 mm range
   (d) Of kernmantle construction and in the 8–10 mm range
9. The term "compatibility" refers to the concept of
   (a) Multiple components being made by the same manufacturer
   (b) Two or more components being well suited to being used together
   (c) Two or more components being the same color
   (d) Inspection methods for equipment
10. Connectors used for rope access should
    (a) Be a self-closing design
    (b) Incorporate a locking mechanism
    (c) Be of sufficient strength for the intended use
    (d) Meet applicable regulations
    (e) All of the above

Answer key:

1. d
2. d
3. e
4. a
5. c
6. b
7. d
8. c
9. b
10. e

KNOWLEDGE CHECK

Chapter 5 Personal Equipment for Rope Access

1. Some of the items typically classified as "personal gear" include
   (a) Ascender, Descender, Cow's tail, Primary system rope
   (b) Descender, Cow's tail, Primary system rope, backup device
   (c) Backup device, cow's tail, descender, ascender
   (d) Backup device, backup system rope, handled ascender, chest ascender
2. Personal protective equipment (PPE) typically includes that equipment which is
   (a) Used for rope access backup
   (b) Worn by the user for protection against hazards
   (c) Owned personally by the rope access technician
   (d) Provided by the employer
3. Any equipment used for life safety must
   (a) Be explicitly designed and intended for life safety use
   (b) Be selected based on its ability to perform adequately for the task at hand
   (c) Meet applicable regulatory requirements in the jurisdiction where it is used
   (d) All of the above
4. A harness used for rope access must
   (a) Have at least dorsal and waist attachments
   (b) Have at least sternal and dorsal attachments
   (c) Have at least waist and sternal attachments
   (d) Have at least waist, sternal, and dorsal attachments
5. A comfort seat may be used for rope access instead of a harness when
   (a) Both the primary and secondary attachments are connected to the same point on the seat
   (b) The user also uses a seat harness for backup
   (c) Technicians on the jobsite are all in agreement
   (d) Never
6. Which selection criteria are important for a helmet used in rope access work?
   (a) Adequate protection
   (b) Comfort
   (c) Inclusion of a three-point retention system
   (d) All of the above
7. A rope access cow's tail is
   (a) A type of lanyard commonly used by a rope access technician
   (b) Constructed of bovine material
   (c) Typically made from dynamic rope
   (d) Both a and c are true
   (e) Both a and b are true
8. A rope access backup device
   (a) Should be easy to put on and take off the rope
   (b) Should be able to function when used in odd directions of travel

(c) May need to be compatible with a rescue load
(d) All of the above

9. Descenders should be selected based on
   (a) Inclusion of autolock, panic lock, and/or adjustable friction features (if desired)
   (b) Mass of intended load
   (c) Compatibility with rope type and diameter
   (d) All of the above

10. A chest ascender
    (a) May be designated as being for either left- or right-handed technicians
    (b) Features connection points designed for rigging into the sternal part of a harness
    (c) Is the same as a handled ascender
    (d) All of the above

Answer key:

1. c
2. b
3. d
4. c
5. d
6. d
7. d
8. d
9. d
10. b

KNOWLEDGE CHECK

Chapter 6 Rigging Concepts

1. When rope access systems are built near the surface of the earth, gravitational force is
   (a) Persistent
   (b) Always a factor
   (c) Equal to the weight of the load
   (d) All of the above
2. A fall line is
   (a) The rope that a technician uses to protect against a fall
   (b) The path a ball would take if rolled down a slope
   (c) A rope with a ball on the end for estimating slope angle
   (d) All of the parts of the backup safety system, rigged together
3. Friction in a rope access system
   (a) Is dangerous and should be avoided at all costs
   (b) Can be used for positive purposes within the system
   (c) Can damage ropes and equipment if not planned for and mitigated properly
   (d) Both b and c are true
4. Acceptable methods for protecting against the effects of friction include
   (a) Edge rollers
   (b) Plastic edge pads
   (c) Canvas edge pads
   (d) None of the above
   (e) All of the above
5. A force vector always has
   (a) Direction
   (b) Magnitude
   (c) A re-direction
   (d) Answers a, b, and c are correct
   (e) Only answers a and b are correct
6. Two important considerations regarding angles in a system are
   (a) Strength loss of material and the descender being used
   (b) The color of the rope and multiplication of forces
   (c) The direction of force and multiplication of forces
   (d) The load ratio and the descender being used
7. Actual Mechanical Advantage is typically
   (a) Less than Theoretical Mechanical Advantage due to friction, angles, and other factors
   (b) Greater than Theoretical Mechanical Advantage due to friction, angles, and other factors
   (c) The same as Theoretical Mechanical Advantage
   (d) None of the above

8. A load ratio is
   (a) A measure of the difference between the strength of something as compared with the load that will be exerted upon it
   (b) The ratio between a Theoretical Mechanical Advantage and the corresponding Actual Mechanical Advantage
   (c) The difference between the safety factor and the mechanical advantage in a system
   (d) The relative proportion of strength retained in a component after it is loaded
9. A safety factor is determined by finding
   (a) The ratio between each individual component and the weight of the load
   (b) The ratio between the weakest point in the system and the maximum potential load that is reasonably foreseeable at that point.
   (c) The ratio between the number of people on the system at any given time and the weakest point in the system
   (d) The ratio between each individual component and the number of people on the system at any given time
10. The target system safety factor should be
    (a) At least 15:1
    (b) Established by a qualified or competent person, based on circumstances and need
    (c) The same for every system, in every situation
    (d) The same as the strength of the rope

Answer key:

1. d
2. b
3. d
4. e
5. e
6. c
7. a
8. a
9. b
10. b

KNOWLEDGE CHECK

Chapter 7 Rope Terminations and Anchorages

1. The running end of a rope is considered to be that part which is
   (a) Used to tie or rig the knot
   (b) A rope in use
   (c) The part of the rope that is left free for use
   (d) Left over in the knotted end of the rope
2. Sewn Terminations in life safety rope
   (a) Are not permitted, ever
   (b) Must be installed by the manufacturer or their designee
   (c) Are appropriate only as a backup to a knot
   (d) Are used to stitch knots in place
3. The difference between a bight and a loop is that
   (a) A loop has a part of the rope crossing itself to form a closed loop, while a bight simply forms a U shape
   (b) A bight has a part of the rope crossing itself to form a closed loop, while a loop simply forms a U shape
   (c) There is no difference between a bight and a loop – these two terms refer to the same thing
   (d) A bight is sewn closed while a loop is knotted
4. Knots are used in rope access to
   (a) Form a loop at the end of a rope
   (b) Form a loop in the middle of a rope
   (c) Connect two ropes
   (d) All of the above
5. An anchorage is
   (a) A type of knot used midline
   (b) A place or fixture that supports and to which the various ropes and rope systems are attached
   (c) The means of securing to an anchorage, such as a strap, eye bolt, etc.
   (d) Also known as an anchorage connector
6. In a rope access system, the primary system and the backup system
   (a) Must share the same anchorage
   (b) May share an anchorage, as long as it is 2× the required strength
   (c) Need not both meet minimum strength requirements
   (d) Must each have their own anchorage
7. Industry best practice for anchorage strengths in the United States calls for at least
   (a) 2,000 lb
   (b) 5,000 lb
   (c) 10,000 lb
   (d) 15,000 lb
8. A backtie is
   (a) A type of anchor system that uses knotted ropes
   (b) A method used to provide security and stability to an anchorage

(c) A type of load sharing anchor
(d) None of the above
9. A load sharing anchor
   (a) Is used to distribute the load between two or more anchorage points
   (b) Is considered to be one anchor system, even when it uses multiple anchorage points
   (c) Is useful when the position or arrangement of a single anchorage point is not sufficient to direct or resist the force as desired
   (d) All of the above
10. An angle of greater than _____ degrees in an anchor system is often referred to as the "critical angle" because beyond this range the forces can become untenable
    (a) 45
    (b) 90
    (c) 120
    (d) 240

Answer key:

1. c
2. b
3. a
4. d
5. b
6. d
7. b
8. b
9. a
10. c

KNOWLEDGE CHECK

Chapter 8 Rope Access Systems

1. The use of rope access methods are appropriate for safety in work at height in
    (a) General industry applications
    (b) Manufacturing plants
    (c) Construction work
    (d) All of the above
2. An effective rope access program for work at height requires
    (a) Suitable equipment
    (b) Worker competence specific to rope access
    (c) Appropriate management
    (d) All of the above
3. A rope access system should be prepared and installed by
    (a) Any rope access technician who is appropriately trained in rigging methods
    (b) Only a Qualified Person
    (c) Only a Level 3 Technician
    (d) Only a Program Manager
4. An Access Zone is
    (a) Limited to the area(s) in which a technician gets on or off the rope
    (b) Any area in which personnel may be at risk of falling, such as on-line or near a working edge
    (c) A city or municipality that permits rope access
    (d) Another name for a rope access jobsite
5. A Hazard Zone includes
    (a) Any area where a worker may be at risk as a result of the work being performed
    (b) Any area where the public may be at risk as a result of the work being performed
    (c) Both a and b
    (d) None of the above
6. The Access System is a system that
    (a) Provides the primary support for the rope access technician
    (b) Provides secondary protection to the technician in the event of failure of the primary system
    (c) Provides both primary and secondary protection to the technician
    (d) Provides security to bystanders
7. The technician may elect to reposition their system by changing the fall line, which may be accomplished by means of a
    (a) Deviation
    (b) Rebelay
    (c) Either a or b
    (d) None of the above

8. A re-anchor system where the subsequent anchorage is set less than 6 ft offset from the original line is known as a
   (a) Short Rebelay
   (b) Long Rebelay
   (c) Mid-rebelay
   (d) None of the above
9. The concept of double protection in rope access refers to
   (a) A requirement that the supervisor must check the technician's harness before they go on a rope
   (b) The simultaneous use of both an access system to support the worker and a backup system to protect the worker in the event of failure
   (c) The fact that rope access technicians should never work alone
   (d) The presence of both a waist and a sternal attachment point on the technician's harness
10. Rope access safety is the responsibility of
    (a) The Program Manager
    (b) The Supervisor
    (c) The technician
    (d) All of the above

Answer key:

1. d
2. d
3. a
4. b
5. c
6. a
7. c
8. a
9. b
10. d

KNOWLEDGE CHECK

Chapter 9 Descending

1. Rope Access descending differs fundamentally from recreational rappelling in
   (a) The simultaneous use of a primary and a backup system
   (b) The training and certification of the technician
   (c) Equipment designed specifically for industrial use
   (d) All of the above
2. An auto-locking descender is one that
   (a) Requires the use of a key
   (b) Locks in place on the rope automatically when released by the technician
   (c) Prevents the technician from descending too rapidly
   (d) Requires three separate and distinct actions to be applied to the rope
3. For a long descent, the negative effects resulting from the weight of the rope beneath the technician may be mitigated by the use of
   (a) A descender with adjustable friction
   (b) A rebelay
   (c) A deviation
   (d) Either a or b
4. Rope access technicians typically use ropes and descenders in the approximate diameter range of
   (a) 7 mm
   (b) 11 mm
   (c) 13 mm
   (d) 16 mm
5. When rigging for descent, the technician should ensure that there are two rope systems (a primary and a backup) rigged so that
   (a) They are closely adjacent to each other and run parallel to the work location
   (b) They are at least 6 ft apart
   (c) They are at least 6 ft apart and run parallel to the work location
   (d) They are closely adjacent to each other and at least 6 ft apart
6. Descenders operate based on the application of friction, which also creates
   (a) Light
   (b) Moisture
   (c) Ice
   (d) Heat
7. Descending with a handled descender is a two-handed procedure, requiring
   (a) One hand to operate the descent control handle while the other holds the control rope as it feeds into the descender from below
   (b) One hand to keep the technician away from the structure while the other operates the descent control handle
   (c) One hand to feed the rope into the descender from below while the other untangles the excess rope
   (d) Both hands to feed the rope into the descender from below

8. During descent, technicians should keep their backup device
   (a) Low, so it doesn't accidentally catch
   (b) High, to minimize potential fall distance in the case of a work positioning line failure
   (c) On their harness, in case they need it
   (d) At the level of the anchor
9. Passing an obstruction in the descent line, such as a knot, requires the technician to
   (a) Swing at least 6 ft from the primary to the backup system
   (b) Momentarily be suspended on only one line, without a backup system
   (c) Perform one, or a series of, standard changeover(s) between descent and ascent
   (d) Flip upside down to release their descender
10. Technicians who intend to perform work while on descent must also be capable of
    (a) Lifting their own body weight on one arm
    (b) Extricating themselves from a stuck predicament
    (c) Ascending out of any situation into which they might get themselves
    (d) Both b and c

Answer key:

1. d
2. b
3. d
4. b
5. a
6. d
7. a
8. b
9. c
10. d

KNOWLEDGE CHECK

Chapter 10 Ascending

1. A technician who is ascending should also employ a secondary/backup system
   (a) At all times
   (b) Only when the primary anchor system is sketchy
   (c) Only when he is passing a deviation
   (d) Only when his primary ascending system is rigged on a 7 mm or less rope
2. A chest ascender and handled ascender
   (a) Must be used together as part of a complete system for ascending
   (b) Must never be used for fall arrest
   (c) Are interchangeable
   (d) Both a and b are true
3. The chest ascender should be rigged
   (a) Above the sternal D-ring
   (b) Between the sternal and waist D-rings
   (c) Below the waist D-ring
   (d) On the back of the harness
4. The handled ascender should
   (a) Be secured to the technician's harness with a cow's tail
   (b) Be rigged with an etrier or footloop for the technician to use as a step
   (c) Never be used for fall arrest
   (d) All of the above
5. The term "changeover" refers to the process of
   (a) Moving from one rope access company to another
   (b) Transitioning from one's primary system to the backup system while suspended on a rope
   (c) Changing from ascent to descent, or from descent to ascent, while suspended on a rope
   (d) Both b and c are true
6. An auto-locking descender may be used to ascend a rope, instead of the chest ascender, and is a good choice
   (a) When the technician wants a more efficient technique
   (b) For very long ascents
   (c) For short distances and when passing obstructions
   (d) When the technician doesn't want to learn how to use a chest ascender
7. When performing a rope-to-rope transfer, technicians must
   (a) Take measures to prevent swing fall
   (b) Ensure that they can reach the second set of ropes
   (c) Use both an ascending system and a descending system simultaneously
   (d) All of the above
8. When negotiating an edge, the technician must
   (a) Maintain adequate protection of the rope(s) from sharp edges
   (b) Use two sets of ropes, for a total of four ropes (two primary, two backup)

(c) Build a rope ladder

(d) Remove the backup protection so it doesn't get in the way

9. When approaching an obstruction from below while on ascent, it is often prudent for technicians to
   (a) Lock off their backup device before negotiating the obstruction
   (b) Perform a changeover just below the obstruction to ensure that they begin the maneuver suspended from their descender
   (c) Wait for the supervisor to watch them while they perform the maneuver
   (d) Tie a butterfly knot just below their chest ascender

10. Rope access technicians who may be ascending as part of their work should be capable of
    (a) Getting themselves out of whatever predicament they might be able to get themselves into
    (b) Disentangling themselves from a stuck system
    (c) Rescuing a coworker who might become incapacitated
    (d) All of the above

Answer key:

1. a
2. d
3. b
4. d
5. c
6. c
7. d
8. a
9. b
10. d

KNOWLEDGE CHECK

Chapter 11 Advanced Techniques

1. Advanced techniques are an important part of the technician's toolbox, as they
   (a) Are fun to do, and therefore add recreation to an otherwise potentially dreary workday
   (b) Allow the technician to show off for coworkers
   (c) Potentially increase pay grade
   (d) Can be key in helping the technician to maintain protection under circumstances that might otherwise be deemed "infeasible" to protect
2. The purpose of a belay in context of rope access system is to
   (a) Catch the load in case of a failure in the primary system
   (b) Stop all forward progress
   (c) Give the supervisor more control over the worker
   (d) Use additional equipment
3. Aid climbing is a method used by rope access technicians to
   (a) Assist one another through difficult maneuvers
   (b) Access an expanse where ropes have not yet been set, and where the structure itself is not conducive to climbing
   (c) Rescue a coworker from a knot pass
   (d) Avoid having to always maintain at least two points of connection
4. Lead climbing is a technique, used only in rare circumstances by specially trained technicians, which allows a technician to
   (a) Work independently, without the need for another technician or coworker on the site
   (b) Perform beyond the level of their training
   (c) Directly climb a structure while protected by an appropriate harness and safety line, belayed by a second technician
   (d) Lead an untrained coworker through rope access maneuvers to access a worksite
5. When climbing with twin lanyards, the technician
   (a) Is potentially exposed to ground impact throughout the course of the climb
   (b) May experience significant impact force if he falls
   (c) Uses the structure as the primary point of contact, and alternately clips and removes lanyards for backup while moving forward
   (d) Is suspended by one leg of the twin lanyard while the other leg serves as backup
6. Raising and lowering systems are useful for
   (a) Positioning a technician who needs both hands free to work
   (b) Rescue
   (c) Positioning a worker who is not rope access certified
   (d) Any of the above
7. A piggybacked haul system is one in which
   (a) A separate haul system is anchored and connected to the primary system with a substantial rope grab

(b) The rope in the primary system is also used as the haul line

(c) The backup system is attached to the primary system and used for hauling

(d) More than four pulleys are used to create the haul system

8. Tensioned rope systems

    (a) May be used horizontally as a highline, or angled as a guide line

    (b) Are particularly susceptible to overloading

    (c) Must be rigged with adequate attention to backup safety

    (d) All of the above

9. Powered assist systems

    (a) May be used for ascending and descending

    (b) May be used for raising and lowering

    (c) May be used for rescue

    (d) All of the above

10. The premise of 100% protection is fundamental to safe and effective rope access

    (a) Except when performing advanced techniques

    (b) Except when performing rescue

    (c) At all times

    (d) Some of the time

Answer key:

1. d
2. a
3. b
4. c
5. c
6. d
7. a
8. d
9. d
10. c

KNOWLEDGE CHECK

Chapter 12 Use of Powered Rope Access Devices

1. Powered devices for rope access offer the advantage(s) of
   (a) A marked reduction in the effort required for ascending
   (b) A stable comfort seat when working on a rope
   (c) Ability to operate as a fixed winch
   (d) All of the above

2. Powered devices should be used for rope access
   (a) Only by competent, certified rope access technicians
   (b) Only by those who also possess additional training relative to the powered equipment they are using
   (c) Both a and b are true
   (d) As a substitute for appropriate training and experience in rope access

3. When using a powered system within the context of typical rope access
   (a) The technician is connected to the primary progress line via the powered device and the technician is also connected to an appropriate backup system
   (b) The rope access technician uses a manual ascending or descending system on his primary progress line, while the powered device serves as backup
   (c) The technician does not need a backup system
   (d) The powered device serves as both the primary and the backup system

4. The optimum number of wraps that the rope should take around the capstan will depend on
   (a) The design and specifications of the device
   (b) How much load will be placed on it
   (c) The condition of the rope and contact surfaces
   (d) All of the above

5. Depending on the design and model of device used the technician may
   (a) Work while sitting on top of (and connected to) the device
   (b) Work while suspended beneath (and connected to) the device
   (c) Be suspended from the rope while the device is operated on the ground by another technician
   (d) Any of the above may be true

6. When using a powered device from a fixed position to raise and lower a technician, a minimum of _____ persons (including the technician being raised/lowered) is required
   (a) 1
   (b) 2
   (c) 3
   (d) 4

7. Powered devices are most effective
   (a) On worksites where the work is mostly vertical (ascending and descending)
   (b) Where extensive maneuvers, such as rope-to-rope transfers, knot passes, or other mid-line actions, are required

(c) Where the working line contains numerous knots
   (d) Where the working line(s) are rigged primarily in horizontal configurations
8. Powered devices should be inspected prior to use to ensure that
   (a) Load-bearing suspension points are not damaged or excessively worn
   (b) Handles, buttons, levers, and rope grabs operate properly
   (c) The capstan drum rotates in the direction it should and not in the opposite direction
   (d) All of the above
9. A hazard analysis for a jobsite where a powered device(s) will be used should include
   (a) Ventilation and hot surfaces
   (b) Entanglement and rescue
   (c) Information specific to the make/model of equipment being used
   (d) All of the above
10. A rope access technician who is trained on one brand/model of powered device
    (a) Should be fine using all brands/models of powered devices
    (b) Should be specifically trained before using any other brand/model of powered device
    (c) Should not be cross-trained on any other brand/model, to avoid confusion
    (d) Should use the device for all jobs from that point forward

Answer key:

1. d
2. c
3. a
4. d
5. d
6. b
7. a
8. d
9. d
10. b

KNOWLEDGE CHECK

Chapter 13 Rescue

1. Employers should preplan for post-fall rescue
   (a) Only when the employees are inexperienced
   (b) Whenever a municipal rescue agency is more than 27 minutes away
   (c) Any time workers are exposed to a potential fall
   (d) When an NFPA certified rescue technician is available
2. A rescue preplan should take into consideration
   (a) Self-Rescue
   (b) Coworker-Assisted Rescue
   (c) Professional Rescue
   (d) All of the above
3. A rope access technician is generally better prepared for self-rescue than a worker using conventional fall protection because
   (a) Of the extensive training he receives
   (b) The use of sternal attachment for fall protection
   (c) The range of equipment that he carries on his harness
   (d) All of the above
4. The guiding principle(s) that a technician should rely upon when selecting which approach of coworker-assisted rescue fits a given situation are
   (a) Simplicity, safety, and efficiency
   (b) How well he likes the affected worker
   (c) Cost, time, and height
   (d) Time of day
5. Noncommittal rescue refers to any method of rescue where
   (a) The rescuer really doesn't want to be there
   (b) The rescuer is able to effect the rescue without entering a fall hazard
   (c) The rescuer is not committed to the outcome
   (d) Ropes are not used
6. A technician who is preparing to perform coworker-assisted rescue from ascent should
   (a) Notify emergency services before commencing the effort
   (b) Work quickly, even if safety is compromised
   (c) Prioritize the safety of their coworker above all else – including his own safety
   (d) Hold his knife in his teeth as he approaches the subject
7. When performing pickoff rescue from descent, the rescuer should
   (a) Avoid attaching the subject directly to the rescuer's harness
   (b) Always attach the subject directly to the rescuers harness
   (c) Never use more than one connection to the subject
   (d) Always use separate descenders for the subject and rescuer
8. Suspension intolerance is
   (a) A potential concern for anyone who has been the subject of a rescue
   (b) Aggravated by motionless suspension

(c) All of the above

(d) None of the above

9. If the potential exists for a rescue scenario to exceed the capabilities of on-site workers, rescue may be achieved by
   (a) Cutting the rope
   (b) Osmosis
   (c) Standby rescue
   (d) None of the above

10. Factors that can make rescue from rope access more challenging include
    (a) High wind
    (b) Extreme height
    (c) Remote location
    (d) All of the above

Answer key:

1. c
2. d
3. d
4. a
5. b
6. a
7. a
8. c
9. c
10. d

Chapters 14–23 consist of worksheets and instructions for developing a rope access program, and therefore do not require knowledge check questions.

Glossary

Anchor System The combined anchorage and anchorage connector, working together to support a load.

Access System (See Also Primary System, Working System) In rope access, the system used to attain access, egress, or support at a place of work.

Access Zone Any area in which personnel may be at risk of falling, such as online or near a working edge.

Active Fall Protection Any harness-based system, such as restraint, rope access, positioning, or fall arrest, which is intended to protect a worker from a potential fall.

Actual Mechanical Advantage True amount of mechanical advantage offered by a haul system after taking into account efficiency loss due to friction, angles, and other factors.

Adjustable Friction Descender A descender with features that allow for addition or reduction of friction to accommodate varying loads.

Administrative Controls Warning methods (such as training, warning signs, lights or sounds) that signal or warn an authorized person to avoid approaching a fall hazard.

Aid Climbing Method used by a technician to access an expanse by placing anchorages along the structure to support equipment (footloops) for progression.

Anchor Point (See Also Anchorage) A place, fixing or fixture, that supports and to which the various ropes and rope systems are attached.

Anchorage (See Also Anchor Point) A place or fixture that supports and to which the various ropes and rope systems are attached.

Anchorage Connector The means of securing to an anchorage, such as a strap, eye bolt, and so on.

Angle The point of direction change between two intersecting lines or surfaces at or close to the point where they meet.

Authorized Person An individual who is approved or assigned by the employer to perform a specific type of duty or duties or to be at a specific location or locations at the jobsite.

Backup System In rope access, the system is used as secondary protection against a fall.

Belay Method used by a technician to stop or catch a load in case of a fall. A "passive belay" activates automatically, while an "active belay" is operated by a person.

Bend Type of tie (knot) that connects two rope ends together.

Bight The shape made when a rope takes a U-turn on itself so that the running end and standing end run parallel to each other.

Bosun's Chair A colloquial term that has been at times used to generically describe rope-based methods of working at height. NOT synonymous with the term "rope access."

Professional Rope Access: A Guide To Working Safely at Height, First Edition. Loui McCurley.
© 2016 John Wiley & Sons, Inc. Published 2016 by John Wiley & Sons, Inc.

Braking Device (See Also Descender) An equipment component used by a rope access technician to apply friction to a rope for the purpose of controlling the rate of speed of descent. May also be used for ascent, hauling equipment, rescue, or other purposes.

Buddy Check Practice of inspecting a coworker's rigged harness and equipment to verify their readiness for work.

Changeover Process used by a technician to change from ascent to descent, or from descent to ascent, while suspended on rope.

Compatibility Appropriateness of any given piece of equipment for the function that it has to perform, especially in relation to adjacent components.

Competent Person One who is capable of identifying existing and predictable hazards in the surroundings or working conditions, which are unsanitary, hazardous, or dangerous to employees, and who has authorization to take prompt corrective measures to eliminate them.

Comprehensive Managed Fall Protection Program A complete, documented plan for provision of fall protection within an organization, including specifications for equipment, systems, and personnel, as well as training, management, and oversite.

Controlled Descent A colloquial term that has been at times used to generically describe rope-based methods of working at height. NOT synonymous with the term "rope access."

Cross-Haul Method used by a technician to move a load in multiple directions at once by using multiple moving rope systems simultaneously.

Descender (See Also Braking Device) A type of braking device designed to apply friction to a rope for the purpose of controlling the rate of speed of descent. May also be used for ascent, hauling equipment, rescue, or other purposes.

Deviation A method of redirecting a line to a different fall line by clipping an access line or a backup line into a midline anchor so that it runs freely through an anchored connector or pulley device.

Double Protection Refers to redundancy within a system to ensure that the failure of any one component will not result in catastrophic failure.

Edge Negotiation Process used by a technician to move past an edge while suspended on rope.

Edge Protection Padding, rollers, or other materials placed on a surface to protect ropes or other equipments against abrasion or damage.

Employer An individual or organization that is obligated under an express or implied contract to compensate another for work or services.

End of Line Knot A knot tied at the end of a rope.

Etrier (See Also Footloop) A flexible ladder, usually consisting of three to six steps, made from webbing or cordage.

Fall Hazard Survey A report that identifies fall hazards as well as ways and means by which the hazards may be eliminated or controlled.

Fall Line The path of least resistance down any particular part of a slope.

Footloop (See Also Etrier) A type of etrier consisting of just one step.

Guide Line A type of tensioned rope system that is rigged in a sloping configuration between two anchors.

Hazard Zone Any area where a person may be at risk (whether by exposure to a fall, falling objects, or work hazards) as a result of the work being performed.

Highline A type of tensioned rope system that is rigged in a horizontal or near horizontal configuration between two anchors.

Hitch A type of tie characterized by being formed around an object and conforming to the shape of that object, and that loses its shape when the object around which it is tied is removed.

Independent (Certification) Freedom from undue personal, professional, or financial influence.

Independent (System) Refers to using two systems simultaneously and in a manner such that each does not influence the other so that the failure of one system will not result in catastrophic failure.

Knot A termination made by tying in a rope, cordage, or webbing.

Knot-Pass Process used by a technician to move seamlessly past a knot or knots on access or backup line(s) without compromising redundant safety.

Lead Climbing Method used by a technician to climb a structure while being belayed.

Life Safety Equipment, systems, or processes relating to the protection of human life.

Line Rope in use.

Load Ratio The difference between the strength of a component or system as compared with the load that will be exerted upon it.

Loop The shape made when a portion of the standing end of a rope crosses over or under the running end.

Lowering Method used by a technician to let down a load from above by using moving ropes.

Lowering Device (See Also Braking Device) A type of braking device designed to apply friction to a rope for the purpose of controlling the rate of speed of a rope as it is used to lower a load.

Mechanical Advantage In rope access, it refers to the use of ropes and pulleys to increase the ratio of pulling force as compared with a load.

Midline Knot A termination made by tying along the path of a rope, cordage, or webbing (not at the ends).

Passive Fall Protection Any means of preventing a fall by separating the authorized person from the fall hazard.

Policy Statement A statement by an employer that outlines the employer's commitment to providing a safe workplace for employees who may be exposed to fall hazards.

Powered Systems In rope access, a mechanical device running on gasoline or battery used to ascend rope.

Prevention (Elimination, Substitution) Any method that eliminates or prevents exposure to a hazard.

Primary System (See Also Access System) In rope access, the system used to attain access, egress, or support at a place of work.

Program Administrator One who is responsible for overseeing the preparation, implementation, and review of the comprehensive managed fall protection program.

Program Manager In context of rope access, the individual who is responsible for oversite of the program, including areas such as equipment, risk management, and record-keeping.

Qualified Person One who, by possession of a recognized degree, certificate, or professional standing, and/or who by extensive knowledge, training, and experience, has successfully demonstrated the ability to solve or resolve problems relating to the subject matter, the work, or the project.

Raising Method used by a technician to lift a load from above by using moving ropes.

Rappel A colloquial term that has been at times used to generically describe rope-based methods of working at height. NOT synonymous with the term "rope access."

Re-anchor See Also Rebelay

Rebelay See Also Re-anchor

Rope Access A method of access that provides the user with the means to safely gain access to, be supported at, and as a means of egress from a high place for the purpose of carrying out work.

Rope Access Supervisor A technician with an advanced level of certification and extensive knowledge, skills, and experience in the use of rope access in the workplace.

Rope Descent Techniques (RDS) A colloquial term that has been at times used to generically describe rope-based methods of working at height. NOT synonymous with the term "rope access."

Rope-to-Rope Transfer Process used by a technician to move from one set of ropes to another while suspended on rope.

Safety Factor The ratio between the weakest point in the system and the maximum potential load that is reasonably foreseeable at that point.

Stopper Knot A termination tied in the running end of a rope to prevent the practitioner from descending off the end.

System Interchangeability In rope access, the concept of the primary system being able to be alternatively used as the backup, and vice versa

Tensioned Rope System Method used by a technician to create a track line for continuous suspension of a load across a span.

Termination (of rope) A loop or blockage made in a rope or webbing by tying, sewing swaging, or other means.

Theoretical Mechanical Advantage Calculated amount of mechanical advantage offered by a haul system with no consideration given to efficiency loss due to friction, angles, or other factors.

Track Line The tensioned line in a guideline or highline system.

Twin Lanyard Climbing Method used by a technician to progress along a structure while using two lanyards for protection, alternately clipping and removing them while moving forward to protect against a fall.

Vector A force having both magnitude and direction.

Working System (See Also Primary System, Access System) In rope access, the system used to attain access, egress, or support at a place of work.

Index

A
access system, 6, 151, 163, 181
access zone, 150, 167
active fall protection, 29, 250
actual mechanical advantage, 107
aerial platforms, 17
aid climbing, 6, 18, 203
anchorage connector, 65, 117, 134
anchor(age) point, 116, 137, 139, 206, 222
anchor system, 115, 135
applications for rope access, 148
ascender, 86, 182–186, 219
ascending, 183, 223
authorized person, 23, 29, 289, 312

B
back tie, 136
backup device, 65, 85, 153, 172
backup system, 13, 153, 200, 253
beam clamp, 66
belay, 84, 200–203, 206, 208, 210
bend, 97, 125–172
bight, 116
bosun's-chair, 11–13, 76
braking device, 65, 81, 200–202
buddy check, 33, 258, 304

C
carabiner, 59–62, 79
certification, 7, 15, 40–47, 53, 89, 134, 297
changeover, 44, 173, 177, 188–197
clearance, 167, 179, 204, 209
clothing, 88
comfort seat, 13, 76, 217
compatibility (of equipment), 13, 56–65, 73, 138, 151
compatibility with conventional methods, 15–18
competent person, 23–25, 31, 112–114, 149, 199, 283
Comprehensive Managed Fall Protection Plan, 21–34, 220, 234, 249
connector, 50, 59–62, 79–80
controlled descent, 11
cost of rope access, 4, 19
co-worker assisted rescue, 14, 135, 235–245, 289–296
cow's tail, 78–79, 85, 186, 203
cranes, 53
cross-haul, 212–213

D
descender, 65, 80–84, 164–166, 170, 189
descending, 151, 163–180, 189, 219
deviation, 7, 44, 143, 157, 175, 194
double protection, 30, 42, 149, 254

E
edge negotiation, 136, 169–171, 193
edge protection, 96–98, 151, 171, 193
elimination, 27, 52, 286
employer, 3, 19, 21–26, 46, 73, 232, 249–314
end of line knot, 121
equipment, 49–90
equipment inspection and care, 33, 59, 89, 228, 303–307
equipment strength, 9, 50–51, 56–61, 65, 78–80, 97–98, 112–113, 131–133, 135

F
fall arrest, 4, 6, 9–10, 13–18, 52, 135, 235, 250–254
fall containment, 17
fall hazard survey, 25–26, 249, 283–287
fall line, 95–96, 139–143, 155–160, 194
footloop, 18, 184–188, 203
friction, 51, 65, 81–83, 93–97, 107, 164–166
fundamentals, 30, 93, 115, 251

G
gloves, 88
gravity, 94–96, 98, 112, 139, 170, 210
guide line, 216

H
hardware, 49–88, 164
harness, 6–16, 27, 74–76, 149, 154, 184–186, 232, 250
harness based fall protection, 8–20, 250–253
hazard zone, 149–151
helmet, 76–77
hierarchy of fall protection, 26–27, 250
highline, 57, 103, 216
hitch, 129–131, 171, 238

I
independent (certification), 41, 45, 53, 297
independent (system), 4–7, 135, 149, 153, 154, 167, 240
International Technical Rescue Symposium (ITRS), 97, 133
ISO 22846, 15, 45–47, 253, 256–257

J
Job Hazard Analysis (JHA), 55, 233, 242, 249, 275–279

K
knot, 9, 58, 115–133, 155
knot pass, 172–174, 192–194
knot strengths, 131

L
lanyard, 15–18, 58, 78–79, 203, 209
lead climbing, 18, 19, 204–208
load ratio, 111–113
load sharing anchor, 68, 139–141
lowering, 209–212, 226

M
man baskets, 17–18
mechanical advantage, 105–110, 212, 242–243
midline knot, 124–125, 155
minimum breaking strength (MBS), 51

N
noncommittal rescue, 238–242

P
passive fall protection, 17, 26–27, 200–202

Professional Rope Access: A Guide To Working Safely at Height, First Edition. Loui McCurley.
© 2016 John Wiley & Sons, Inc. Published 2016 by John Wiley & Sons, Inc.

personal equipment, 52, 71–90
personal protective equipment (PPE), 52, 71–73
personnel responsibilities, 23–26, 38–40
personnel selection, 32, 35–47
personnel training records, 41–42, 249, 297–301
policy statement, 22, 249, 259–265
portable high directional, 66
positioning, 15, 28, 135, 227, 250–254
powered systems, 217, 219–228
PPE (Personal Protective Equipment), 52, 71–73
pre-rigging for rescue, 168, 238–240
prevention, 26–27 *see also* elimination; substitution
primary system, 6, 151, 163, 181
professional rescue, 55, 65, 234, 245, 291–296
program administration, 23, 247–314
program administrator, 23–26, 299–308
program audit, 249, 309–314
pulley, 51, 63–64, 97, 100–110, 157
pull-through system, 154–156, 204

Q
qualified person, 23–24

R
raising, 107–110, 209–213, 226–227, 239
rappel, 4, 8–9, 116, 163
Reanchor, 144, 158–159, 176–179, 194–198
rebelay, 144, 158–159, 176–179, 194–198
recordkeeping, 89, 305, 313–314
regulatory requirements, 22, 23, 50–53, 71–73, 135, 166, 232
removable bolt, 66–68, 134
rescue, 24, 32, 54–58, 135, 229–244, 253–254, 289–296
rescue equipment, 54–56, 65, 73
rescue preplan, 234, 245, 254, 289–296
responsibilities, program manager, 23, 40, 257
responsibilities, supervisor, 31–34, 39–40, 244, 257
responsibilities, technician, 7, 23–26, 37–47
restraint, 15, 28, 53, 135, 250
rigging equipment, 49–69
rope, 55–59, 115–118, 166
Rope Access Permit, 2, 249, 256 *see also* Work Plan
rope access supervisor, 2, 257
Rope Access Work Plan, 22, 29, 249, 257, 271–275 *see also* Permit
rope descent techniques (RDS), 11, 30
rope grab, 64–65, 84–86, 182
rope-to-rope transfer, 190
running end of rope, 116

S
safety factor, 51, 111–114, 135
safe working load (SWL), 51, 216
screwlink, 59, 61–62, 79
seatboard, 13, 76, 217
self-rescue, 32, 233–235, 291–293
snaphook, 59, 61
standby rescue, 244, 254, 291
standing part of rope, 116, 124
stopper knot, 119–121
substitution, 27, 52, 286
suspension intolerance, 230–232, 238–241
swing fall, 143, 151, 157–158, 191, 194, 206

T
tail of rope, 83, 116
team organization, 19, 31, 37–40, 255
technician skills, 7, 38–39, 46, 91–246, 301
tensioned ropes, 213–216
termination (of rope), 58, 118–133
Theoretical Mechanical Advantage (TMA), 107–109
track line, 216
training outline, 41–45
twin lanyard climbing, 209

V
vector, 98–104, 136, 143, 213

W
welfare of the technician, 159, 256
wind, 19, 160, 167–168, 172
working end (of rope), 102, 116, 139, 155, 226
working load limit (WLL), 51, 61
working system, 6, 151, 163, 181
work order, 249, 267–270

Printed in the United States
By Bookmasters

small-angle X-ray scattering, 77
smashing limb, 213
soleus, 375
solubility product constant, 129
spatulae, 35, 455, 457–458
spectrin, 114
spherulitic crystal, 176, 216
spider, 88, 296
spider web, 301
Spiderman, 459
spinal cage, 471
spinal implant, 471
spinal plate, 471
spinnerette, 554
sponge, 400
sponge spicule, 160–164
sputtering, 515
squid, 89
squid beak, 339–343
SR, *see* silicon rubber
standard model, 41
STAR scaffold, 555
starling, 437
Starr–Edwards valve, 549
steel, 69
Stegostoma fasciatum, 489
Stenocara sp., 511
stenosis, 548
stent, 385
sterol, 95
stratum compactum, 372
stratum corneum, 362
stress relaxation, 40
stress–strain curve, 301
Strombus gigas, 196, 198
strut, 418
Sturnis vulgaris, 437, 478
sucker, 342
suction, 453
Super Glue, 318
superhydrophobicity, 473
suture, 172, 279
synchrotron, 78
synchrotron-ray computed tomography, 251
synthetic foam, 441
synthetic skin, 576
systolic heart cycle, 378

T4 phage, 563
teeth, 262
Teilhard de Chardin, P., 30
telson, 215
tendon, 293
tensegrity, 108
terraced cone, 179
terrapin, 278
testudine, 278–281
Thalassiosira pseudonana, 158
Thomas tetrahedron, 7
three-dimensional printing, 409
threonine, 58
thymine, 55

tissue engineering, 553–554
tissue engineering scaffold, 553–554
titanium oxide, 270
titanium scaffold, 409
titin, 75
TKA, *see* total knee arthroplasty
toad, 372
Toco toucan, 423, 441
toe pad, 461
torsion, 399
tortoise, 278
total hip replacement, 63
total knee arthroplasty, 65
toucan, 421
toucan beak, 330, 425
trabeculae, 422, 429
transfer RNA, 55
transforming growth factor, 410
transgenic mice, 570
transverse fracture, 240
trap-jaw ant, 344
tree, 33
tree frog, 461
Treloar equation, 359, 390; *see also* Flory–Treloar equation
tricuspid valve, 548
Tridacna gigas, 202
triglyceride, 95
trilobite, 304
triple helix, 69
Troypeutes tricinctus, 274
tryptophan, 58
tubule, 263, 266, 307, 320–321, 324, 327
turbine blade, 546
turtle, 278
tusk, 262
twisted plywood structure, 311, 338
type I collagen, 333
tyrosine, 58

UltraCane, 527
ultrahigh-density polyethylene, 63
ultrahigh-molecular-weight polyethylene, 63
uncracked ligament, 246
underwater adhesion, 465
urinary incontinence, 552

vacuole, 105
vaginal tissue, 79
valine, 58
van der Waals attraction, 455, 460, 464, 470, 512
vascular channel, 224
vascular implant, 387
vascular system, 378
vascularization, 577
vaterite, 146, 217
vein, 378
VELCRO®, 453, 497
vena cava, 548
venous blood, 547
vent hole, 102

ventricle, 548
venule, 378
Venus flower basket, 161
vesicle, 555
virus, 563, 565, 570
virus–DNA array, 498
viscid silk, 297
viscoelastic, 323
viscoelastic behavior, 369
viscoelasticity, 40
Vitallium, 5, 172
Voigt model, 226, 325
Volkmann's canals, 224, 256
Von Lilienthal, O., 497
vulture wing, 505

water, 54–55
water collection, 511–512
water-jet mechanism, 340
wave-guiding properties, 509
wax, 95
Weibull, W., 45
Weibull equation, 47
Weibull modulus, 46
Weibull statistics, 186, 206
wet adhesion, 453

wet spinning, 581
whale bulla, 274
whale fin, 546
whelk egg, 387–390
wing bone, 400
wing case, 315
WLC model, see worm-like chain model
Wolff's law, 33, 224
wood, 400, 410–417
wool, 390
work of fracture, 184
worm-like chain equation, 355
worm-like chain model, 115, 355
woven composite, 539
Wulff plot, 132

X-ray diffraction, 560
xanthophore, 483

Z line, 85
Zayed National Museum, 508
zebra shark, 489
zinc phosphate, 318
zirconia, 142
zooplankton, 160

Printed in the United States
By Bookmasters